Common Circuits

Hacking Alternative Technological Futures

LUIS FELIPE R. MURILLO

STANFORD UNIVERSITY PRESS
Stanford, California

Stanford University Press
Stanford, California

Printed in the United States of America on acid-free, archival-quality paper.

Library of Congress Cataloging-in-Publication Data
Names: R. Murillo, Luis Felipe, author.
Title: Common circuits : hacking alternative technological futures / Luis Felipe R. Murillo.
Description: Stanford, California : Stanford University Press, 2025. | Includes bibliographical references and index.
Identifiers: LCCN 2024028512 (print) | LCCN 2024028513 (ebook) | ISBN 9781503640603 (cloth) | ISBN 9781503641488 (paperback) | ISBN 9781503641495 (ebook)
Subjects: LCSH: Information technology—Social aspects. | Information technology—Political aspects. | Hacking—Social aspects. | Hacking—Political aspects. | Hackers. | Cooperation.
Classification: LCC HM851 .M866 2025 (print) | LCC HM851 (ebook) | DDC 303.48/33—dc23/eng/20240626
LC record available at https://lccn.loc.gov/2024028512
LC ebook record available at https://lccn.loc.gov/2024028513

Cover design: Bob Aufuldish, Aufuldish & Warinner
Cover art: A64-OLinuXino board design courtesy of Olimex, generated by the author with KiCad

Common
Circuits

Au fond, ce sont des mélanges. On mêle les âmes dans les choses; on mêle les choses dans les âmes. On mêle les vies et voilà comment les personnes et les choses mêlées sortent chacune de sa sphère et se mêlent: ce qui est précisément le contrat et l'échange

MARCEL MAUSS, *Essai sur le don*

Contents

Illustrations

Acknowledgments

This book is a product of several reference circuits, peopled with those who realize impossible dreams of positive social and technical change. It is to them that I dedicate this volume. Educators and technologists, hackers and non-hackers alike: they are all virtually present here through the gifts that I have received from them. I thank Carlos Steil for the gift of anthropology; Isabel Carvalho for being my partner in the cosmopolitical flights of scholarly imagination; Chris Kelty for being an unparalleled mentor and guide through the common(s); Sharon Traweek for mentoring not only me but several generations on the importance of collaborative scholarship in science and technology studies; Gabriella Coleman for everything she has done and continues to do for us in anthropology; Michael M. J. Fischer for being a constant, positive force for the realization of this book; Michel Lallement for providing an academic home away from home in addition to all the lessons in the sociology of communitarian life; Mariko Tamanoi for patiently guiding me through the challenges and joys of field research in East Asia; Yan Yunxiang for everything he taught us in economic anthropology within and beyond the mainland; Maria Cristina Leandro Ferreira for showing us the importance of language, history, and memory in everything we do; José Carlos dos Anjos for bringing me to the Quilombola territory where I received my first practical lesson on the transformative potential of gift relations; Sareeta Amrute for thinking with me about computing from major/minor circuits; Nelson and Kathy

Graburn for their unwavering support across the continent; Erin McElroy
for the being such a good colleague and friend for everything technical
and political; Leila Adu-Gilmore for all the mighty (*irie*) energy when it
was most needed; Walter Scheirer for our great partnership in the archae-
ology of hacking; Cedric Honnet for our nomadic wanderings in search for
community computing; Susan Blum for her teachings in the anthropology of learning; Héctor Beltrán for being my *compa* in the study of computing across borders; Phil Bourne for defending the common in science
and technology where nobody else does; the Pardos (Luis, Sharmon, Sol,
and Paz) for their luminous presence in computing, music, literature, and
critical legal work; Jarrett Zigon for supporting my strange adventures at
the intersection of anthropology and computing. Many thanks and praises
to my patient and loving kin: Regina, Uirajá, Léo, Allegra, Brian, Linda,
Lulua, Stella, Maria, Paulo, Lucas, Fernanda, and Antônio. Special thanks
to Aidan Seale-Feldman, whose love has always grounded me through
every site and every page, and Rosa Maris Rosado for opening the path of
human sciences and computing for me early on.

Much of what I have learned about anthropology, I owe to my professors and colleagues at the Federal University of Rio Grande do Sul
(UFRGS) and University of California, Los Angeles (UCLA). In particular, I would like to thank Ceres Victora, Bernardo Lewgoy, Ondina Fachel
Leal, Caleb Faria, Ruben Oliven, Arlei Damo, Cornelia Eckert, Mauro Mes-
sina, Sherry Ortner, Akhil Gupta, Jarita Holbrook, Sam Weeks, Liz Collins,
Jason Throop, Bob Kurik, Tereza Virtová, Ann Walters, Leah Lievrouw,
and Ramesh Srinivasan.

I am extremely grateful for all the support I received from my colleagues
and friends at the University of Notre Dame, especially to Insha Bint Bashir,
Jessica Ashman, and Yun Xie for their help with the book manuscript. A
huge thanks to my colleagues at the Institut Francilien Recherche Inno-
vation Société (IFRIS), Laboratoire Interdisciplinaire pour la Sociologie
Économique (LISE-CNRS), École des Hautes Études en Sciences Sociales
(EHESS), and Centre Maurice Halbwach (CMH) as well: Rigas Arvanitis,
Pierre-Benoît Joly, Marie-Christine Bureau, Isabelle Berrebi-Hoffmann,
Christian Azaïs, Olivier Giraud, Audrey Lefèvre, Lea Lima, Zaëra Mari-
aux, Ferruccio Ricciardi, Olivier Alexandre, Florence Weber, Benoît de
L'Estoile, Emmanuel Didier, Monique Dagnaud, and Alexis Spire. Special
thanks to the Berkman Center fellows at Harvard for their support in the

earliest stages of development of this book: Sara Watson, Nate Matias, Neal Cohen, Rajesh Veeraraghavan, and Bruce Schneier.

My debt to the hacker community is equally immense. In particular, I would like to thank Niibe Yutaka for all the lessons on the nature of computing power; Mitch Altman for sharing multiple space-times of reflection on computing and community; and Nala for realizing dreams of community building under highly adversarial conditions. Many of my haxen/haxor friends are integral to this book project: Liz Henry (an infinite loop of thanks), Liz Cole, Scott Beibin, Gustavo Gus, Paul Schweizer, TC Silva, Gus Pradella, Ethan Crawford, Rachel (RIP), Seth Schoen, Jake Spaz, Gnusosa, Andy, Miloh, Danny, Rolf, Valerie, Acathla, Alice, Barzi, Audrey, Gabriel, Akiba, MRE, JAM, Sean, Pieter, Taylan, Michelle Poon, Bill Liang, Hao, Terry, Eric Pan, Atommann, Tully, Aqua, Lyman, Devon, Yoyo, Takanori, Hideki, Sayuri, Taka, Junko, and Sheano. Special thanks to Lucas Freeman for thinking with me throughout and to Silvio Carlos for everything he has done for the cause of common computing. To each and every one of you, I am wholeheartedly grateful.

I thank my editor Dylan Kyung-lim White for his amazing editorial guidance and the whole staff at Stanford University Press for their inestimable help. Last but never least, I thank the anonymous reviewers who helped me improve this book in substantial ways.

The research for this book was funded by the University of California Pacific Rim Research Program and the UCLA Terasaki Center for Japanese Studies. I am very grateful for their support. Excerpts from chapters 1 and 3 appeared in *American Anthropologist*. A version of chapter 5 appeared in *Anthropological Quarterly*. Funding for the preparation and open access release of this book was provided by the National Science Foundation (grant #2226425).

Preface

00001010 Experiences with "Hacking"

0.

"Sophisticated attack" is the expression of choice to describe something that could not have possibly been orchestrated by a small group of independent computer experts. Everyone is asking who is responsible, who in the world could have such capacity. Journalists are on the case and they seem to have some evidence, although it is far from conclusive. They exasperatedly declare, "We are witnessing a new era of cyberattacks on computer systems. Everyone and everything is a potential target." In unison they cry, "No system is safe!" Better-informed industry experts tell journalists they have not seen anything like it either. No longer is it a matter of a reactive body politic needing to make an example out of "computer whiz kids" or hacktivists; the stakes are much higher now. They will attempt to bring down politicians. They might even define elections and determine the future of warring nations in heated but veiled geopolitical conflict. And all of this through the new practice of state-sponsored *hacking* in the dismal scenario of a so-called *cyberwar*.

1.

-1d4y w4r3z, od4y, kod3z, virii, m4y3m phor l4m3rz, phun phor h4krz! [Anti-sec] k4m3 4z th3 w4k3 uP c4ll phor th4 1nDuzTry. nobody r34lly

knowz 1f skr1pt-k1dz || el8 h4krz w3r3 th4 smok1n gunZ b3h1nd pr3v10uz
r3l34s3z, bUT th3 st4k3z 1s h1gh now. t3kn1ks 4r3 mor soph1st1c4d,
m1ss1nform4t10n c4mp41gnz too! d3sp1t3 1tz ph4m3 4nd glory phoR
$ecUr1tY, [OpenBSD] w4s cl41m3d to b kompr0m1s3d. th3 xploit 4ll3g-
3dly 4llow3d phor 4n 4tt4cker to kr4sh 4n obsd box 4nd h4v3 pr1v1lg3d
4cc3ss to DDB dur1ng kr4sh r3cov3ry—wh1ch m34nz xc3ss to th3 k3rn3l
1tzs3lf. [RLoxley] from [Team Hackphreak] r3l3as3d [krnl-DoS.c]: 4 PoC
phor xplo1t1n 4 UVM bug on obsd. M4y3m unl34sh3d ... 4s 1t 4lw4yz do3z
wh3n 1t com3z to th3 3ncont3r btw33n th4 [hacker underground] 4nd th3
[infosec] 1ndu$try. 1Nt3r3st1ngly, th3 obsd cor3 d3v t34m could !not k4r3
l33s ... f4k3 xploitz 3v3rywh3r3! #define CRASH_FILE "./forKbomB"

2.

At Beijing Capital Airport (PEK), an engineering management professor
from a very prestigious Chinese university excitedly awaits the arrival of
scores of hackers from various community spaces in North America and
Western Europe. Holding up a colorful sign with a purposefully controversial imprint, "Welcome *Hackers* to China," he is more than ready to
serve as a generous host and introduce foreign engineers, developers, and
artists to a robust network of Chinese academics, administrators, politicians, investors, and industrialists, bringing to the mainland, under the
controversial symbol of *hacking*, the promise of "digital innovation" as a
newfound key to prosperity.

3.

"There are four types of people in the world," we are told by a self-righteous
software developer who managed to implement a massive advertisement
empire on PHP to capitalize on people's relationships: "those who know
and do, those who know but don't do, those who don't know but do, and,
finally, those who don't know and don't do." *Hackers,* he posits with that
unshakable confidence that only comes with an equal amount of naivete,
are "doers." And in one fell swoop, a Silicon Valley corporation popular-
izes a definition of *hacking* in which pretty much everything fits, from the
creation of stone tools onward in the *longue durée* of human history. *Doers*
are, simply put, *hackers.*

4.

Development experts gathering in a boardroom with a comfortable view of Lake Geneva are wondering how to increase public engagement with their humanitarian projects. Some of them have been around long enough to know that many of the tricks and gimmicks have already been tried to "lift"—as they would often say, abusing the topological metaphor of the social pyramid—and to "empower" people in "developing countries" with digital technologies. The least experienced expert in the room tries his luck by coming up with an idea that is, to his surprise, promptly embraced by the senior officers: "Why don't we try a hackathon?" It takes only a few emails and calls, and within a week many IT corporations are enlisted as sponsors, dates and location set, volunteers enrolled. Posters, stickers, T-shirts, pizzas, soft drinks, and cash prizes will be handed out on the big day for *hacking* humanitarian aid.

5.

In the wake of a triple disaster, a Tokyo-based computer club is effervescent, bubbling with people and projects seeking to aid in disaster relief. The open house this Tuesday evening is somewhat special. Keys are being given as a symbolic act to an honorary member of the hackerspace, an internet entrepreneur, digital commons activist, and university administrator who identifies himself proudly as a *hacker of bureaucracies*. He is treated with due reverence for his support of community projects. Hackerspace members do not have to admit how much they know of his capacity to mobilize resources to make any of their projects surf and swell. When resources are obtained from the people on the wrong side of the force, however, the whole community will turn against him.

6.

A feminist collective is adopting 3-D-printing-based fabrication with a political approach to their bodies and destinies. They are located in their lab at a so-called post-utopian cooperative that finds its roots in the anarchist tradition of insurgent Catalonia to design their own 3-D-printed gynecology tools. *Feminist hackers* is how they identify themselves in order to combat the dominant male gaze and their apparatuses from the sphere of women's health. For more than justifiable and urgent reasons, they have made the news to circulate the decades-old and yet invisibilized category of the *feminist hacker*.

7.

The Brazilian national chamber of deputies is the location of an unlikely space, a *hacker* laboratory created under resolution n. 49 (2013) of the internal revenue code and tasked with the promotion of a network between "parliamentarians, hackers, and civil society," animating *hackdays* and *hacker marathons* to create a culture of transparency in the government. Their projects involve displaying public expenditure information using digital technologies for data visualization. They identify themselves with the newfound figure of the *hacker citizen*.

8.

Across the Atlantic, an engineer is eager to make the news as he takes a leap of faith. The setting is someone's office in an undisclosed location. Aseptic measures are in place, but the operation is nonetheless gruesome to witness. A slightly oversized battery pack with a small microcontroller is being inserted and then sutured under his skin. He is proud to call himself a *biohacker*, someone who is willing to *hack* biology as computer enthusiasts *hack* digital systems, exploring biomedical unknowns with a bricoleur approach. The *biohacker* community cannot side with him on this procedure, but he is, unfortunately for all, the one who is getting all the press.

9.

In the kitchen of a community space in the San Francisco Bay Area, a group of artists, housing activists, and former engineers on welfare (laid off due to the dot-com collapse) gather around the symbol of *food hacking* to experiment with all sorts of techniques and ingredients. When asked why an anarchist writer of erotic science fiction would be interested in joining a space dedicated to computer hacking, he proclaims loud and clear for all to hear: "I hack society!"

Common
Circuits

Circuits in Common

It began as a prank. It became a passion and obsession, a crime and stigma, an identity to be defended with pride, an international community, a fad, and finally, a political practice. In the span of six decades of digital experimentation, *hacking* has shape-shifted constantly to become a catch-all term for pretty much anything tentatively clever or subversive or make-shift with varying degrees of technical prowess. Now it seems to be every-where but with radically distinct contours. From cryptic and clever forms of *détournement* to its dissemination as an identifier for disparate practices, *hacking* has traveled far and wide as a practical symbol.

What does this obsession with *hacking* tell us? It was only with the spectacularization of underground computer collectives in the 1980s that hackers made their way into the commercial media spotlight, which brought their practices, if somewhat distorted, into public view. While ambiguously representing "menace" and "solution," hackers have figured ever since in conflicting narratives as rebels and freedom fighters, trick-sters and occultist magicians, cybercriminals and government allies, in-formation security consultants and anti-corporate technologists. Across a fairly wide range of historical experiences, the moral pendulum has oscillated between the identification of hacking with criminal activities and virtuous practices of community-making. But it is beneath the surface

1

of public discourse that we find minor histories of computing connected through a common thread: hackers have stubbornly fostered the *common*[1] as a principle to create *community with, for, and through computing* and have increasingly done so beyond and in connection with Euro-American computerdom.

Common Circuits offers an anthropological study of a transnational network of community spaces that distinguish themselves through their collaborative work-arounds to create alternatives to "Big Tech" computing. Their alternatives are made possible through practices and infrastructures of commoning to redress social and environmental problems. Their ways of making in collaboration demonstrate that "another possible is possible," an expression made memorable by anthropologist Arturo Escobar in his reformulation of the alterglobalization slogan.[2] Against the Big Tech Midas that transforms everything it touches into property, in minor histories of hacking we find explorations of computing for the purposes of community building. In this context, *commoning* refers to the practices for creating, sharing, and sustaining technologies of common interest— where "interest" is not to be taken in its economic or colloquial sense but by its etymological roots: *inter-esse*, an expression in Latin that stands for the "in-between," a social tie and, for our purposes here, a bundle of ties between us and our technical objects.[3] This insistence on sociotechnical ties is an important challenge to the entrepreneurial history of computing for demonstrating that its purported innovations came from elsewhere, through the substantial work of technologists made invisible.[4] The computer enthusiasts who figure in this book have dedicated their lives to (informal) learning through (convivial) sharing, not necessarily by breaking computer security and exploiting remote networks. They have defended the need for community in a world where digital technologies have been designed and implemented to deny one's ability to learn through collaborative exploration.

Much ink has been spilled in the examination of the negative effects of digital computing as supporting infrastructure of surveillance states, where hacking figures as a wild card in not only advancing but also interrupting circuits of ever-expanding dispossession and extraction in neoliberal circuits. In this book, we take a detour into alternative spaces where marginal projects counter dominant circuits with computing as convivial technology.[5] Our starting point is not only that we can better devise alternative technopolitics through minor practices of computing, but also that contemporary hacker sensibilities, spaces, and projects open up fundamental questions regarding the persistent divides between the technical, the ethical, and the social in science and technology.[6] To explore the implications of this debate from the perspective of a group of (mostly) invisible technologists, I pursue counterexamples of conviviality in common circuits[7] of community spaces called "hackerspaces." In this particular context, *common circuits* refer to the infrastructural conditions for informal learning based on open technologies.[8] It is through active participation in these circuits that I describe (1) how technologists become hackers (through a process we call *personification* in anthropology[9]); (2) how hacker projects are *spatialized* through the creation of convivial places for community gathering; and (3) how these two interrelated processes culminate in the *politicization* of computing expertise at transnational scales. As I explain in what follows, these three analytic frames represent the core components of the book.

Common in Computing

From efforts in the 1950s to break free from the caste of computer priests to the cross-pollination of the countercultural movements in the 1960s and 1970s, we find key experiences through which phone and computer networks came to be used as a means for not only war and bureaucratic governance but also community-making. From the first compilers and programming languages to computer hardware kits and software-sharing collectives, early discourses and practices of liberation constitute one of the aspirational legacies of contemporary hacker circuits. To probe for origin stories, one could start in many places, anywhere from the assumed beginnings with historical fragments of Al-Khwarizmi's work on algorithms in Baghdad around 820 CE; René Carmille and his *détournement* of Nazi punch cards and IBM sorting machines during the occupation; or Britain's secret code-breaking Bletchley Park "women computers" and the trials and tribulations of Alan Turing. For our purposes, however, it is more urgent to begin with what anthropologist Michael M. J. Fischer has called the "cultural switching points"[10] through which digital technologies are peopled in alternative circuits. It is by engaging computing *as people mattered* that we find technologies at the scale of everyday human and computer interactions. This approach, we must say, does not entail anthropomorphizing computers (as thinking machines) or reducing social matters to computational problems that have become trillion-dollar industries. Rather, what we need in order to break out of social and technical divides is to study (and foster) new forms of exchange to enable openings for the common between (and beyond) public and property relations. We can say to this effect with a good degree of methodological confidence that every starting point is arbitrary, but some are meaningful.

One of the well-known figures in the folklore of computerdom is computer scientist and mathematician Grace Hopper, an inescapable character, we could say, for fulfilling institutional and technical roles. Her often-quoted piece of wisdom "It is easier to ask for forgiveness than it is to get permission" would become an article of faith for hackers since it came to print.[11] Hers is a tale of intimacy with computers and collaboration with fellow technologies that many hackers would identify as their own. Hopper's biographer, Kurt Beyer, observed that the celebration of "Amazing Grace" has led, unexpectedly, to a misinterpretation of her contributions. Rather than rebellious, Beyer notes, her work was defined by her collaborative and gregarious nature. Hers was an incredible capacity to overcome gender divides and break into the high priesthood of computing in the 1940s.[12] Despite emerging disputes over patents and prior art concerning early computer technologies, informal sharing was common at the time. Early circuits of information exchange were sustained by the circulation of technologists working on similar problems, but they were hardly open by any stretch of imagination. Women mathematicians had to work hard around established men-identified engineers with technical work-arounds of their own. Hopper, for example, gifted engineers with compiler programs that liberated them from having to use machine code to instruct commercial and military computers. They could finally use "natural languages" for creating a common pool of computer programs and calculate, among other "domestic" applications, precise bomb trajectories (including atomic ones). Software was not yet seen as an "original expression" or "innovation" in the state of the art to be enclosed, respectively, with copyright titles or patents.[13]

Another inescapable point of passage in hacker histories is the one that marks the transformation of the technopolitical landscape of computerdom with the conversion of a disability into a special ability. By the end of the 1950s, Joe "Joybubbles" Engressia is known to have come across an important finding that granted him the honorary place of pioneer in the art of exploration of telephony known as "phone phreaking." Joybubbles was born blind but gifted in his ability to hear, memorize, and reproduce sounds within a wide range of frequencies by whistling. Since childhood he had an obsession with the telephone as a medium of exploration, where he sought refuge from painful memories of domestic violence and sexual abuse. Joybubbles is known for having figured out that by whistling a 2,600-hertz tone, he could unlock long-distance calls. This discovery was also made by electronics hobbyists who proceeded to build a device to substitute for JoJo's whistling, the famous Blue Box, an exciting gift widely circulated among phreaks to generate the standard and maintenance tones of the phone system. In his comprehensive history of phone hacking, engineer Phil Lapsley calls our attention to one of the fundamental aspects of this scene: the communal experience in which the telephone came to represent the very embodiment of curiosity as the network was transformed into a site for exploration and discovery, as well as community-based learning.[14]

Phone phreaking is of particular interest to us for pinpointing one of the earliest hacker circuits through which technical knowledge and autonomist values started to converge to form a network-shaped commons. As noted by legal scholar Fred Shapiro, one of the first negative occurrences of the term *hacking* as a "computer-mediated crime" is intimately linked with phreaking. By 1963 the MIT student newspaper *The Tech* published a front-page article titled "Telephone Hackers Active, Services Curtailed." It described the usage of MIT's PDP-1 computer for "war-dialing," or traversing lists of phone numbers to identify modems on the other side of the line. The professor who was responsible for the phone system is alleged to have said, "We do not have too much trouble with the boys; we appreciate their curiosity." Commenting on the case, the article's author also quotes an "accomplished hacker" who voiced a shared sentiment at the time: "The field is open for experimentation." Indeed, it was all open for experimentation within the parameters of exclusivity through which the boys exercised their privilege.[15]

One year before the somber event of his assassination in 1968, Dr. Martin Luther King Jr. set the tone for the upcoming generation of hackers who inherited the countercultural attitude of the phone phreaks. "We must rapidly begin the shift from a thing-oriented society to a person-oriented society," Dr. King urged while signaling to the space where commoning thrives. "When machines and computers, profit motives and property rights are considered more important than people," he said, "the giant triplets of racism, extreme materialism, and militarism are incapable of being conquered."[16] As we reached the peak effervescence of the late 1960s, phreaking was taken further and beyond (gendered) technical exploration and youth pranksterism. The hippie movement was radicalized by the yippies (with their politics illustrated by their flag: a black background with a red communist star and a green cannabis leaf)—and new types of restricted expert knowledge were put into circulation and promptly included in the hacker toolbox. As part of a whole series of circumvention techniques for liberating things from their proprietary modes of existence, lock picking was taught alongside numerous techniques of phreaking in the Youth International Party Line (YIPL). An important newsletter of the movement, which led in the subsequent decade to the formation of the long-lasting *2600 Hacker Quarterly*, YIPL reclaimed phone hacking with an overt political orientation. It represented not only a community for the insatiably curious and technically inclined but, most importantly, a venue for engaging the antiestablishment. Jason Scott, an organic historian of the computer underground, remarks that the newsletter stems from the integrated circuit of radicalized elements of the counterculture: the autonomist New Left of the 1960s.

The wild shake-up in the 1960s would prove pivotal for convivial experimentation and critique of computing as one of the key symbols of bureaucratic and instrumental governance. Computing was still infrastructural, but it would soon become a locus of sociotechnical change with the extension of information, knowledge, and power to nonexperts outside academic and governmental agencies.[17] Media scholar Charlton McIlwain reminds us that before the 1980s, there were only two places nonexperts could have access to computers: research laboratories and workplaces. Yet, neither of these spaces were welcoming to Black and Indigenous students and professionals.[18] Curiously, these places would eventually breed a culture of laser-focused computer enthusiasts with unprecedented degrees of openness toward information sharing among themselves. It was a generation of undisciplined technologists obsessed with technical puzzles but also quite oblivious to the massive industrial-government-military complex—as historian Nathan Ensmenger describes in his historical reconstruction[19]—where they found the conditions to cultivate skills by circumventing restrictions to mainframe and mini computers. As early as 1976, computer scientist Joseph Weizenbaum would identify hackers as an "international phenomenon," the so-called computer bums.[20]

Phone and computer hacking would find fertile ground in communal experiments of the mid-1970s. These experiments of a social and technical nature became well-known for popularizing new ways of imagining how communication technologies could also be designed (within limits) for convivial purposes. Instances of commoning in this period included "homebrew" projects to convert computer networks into autonomous means for exchanging information of local interest.[21] Following social critic Ivan Illich's plan for deschooling society with distributed "learning webs," one famous project involved the creation of a small network of public terminals in the San Francisco Bay Area called Community Memory. As a "peer-matching network," it was rather simple: users identified themselves "by name and address and describe[d] the activity for which [they] sought a peer. A computer would send [them] back the names and addresses of all those who had inserted the same description."[22]

Another minor project with major consequences consisted in imagining and implementing a "computerized bulletin board" for exchanging messages between members of the Chicago Area Computing Hobbyist Exchange (CACHE). In his prehistory of social media, communication scholar Kevin Driscoll dubbed these experiments "improvised workshops" where social technologies were first devised.[23] These workshops enabled outward-expanding circuits where parts, kits, manuals, tricks, tapes, and program listings circulated as gifts to create ties, but they would soon be challenged with commercial enclosures in a not-so-distant future. Think of what the Apple I computer kit would become as soon as it abandoned its hobbyist affordances to serve as a launchpad for a soon-to-be global business. Commoning was made possible through the diversion of chips and parts from the military-patronized local electronic and silicon components industry but also by refocusing expertise from military-industrial affairs to an uncharted experimental, technopolitical terrain.[24] From industry and university-backed digital infrastructures to convivial spaces and back, reciprocity ties would break as hacker technologies left their common circuits to enter the register of the nascent personal computer market. The "personal computer revolution," anthropologist of technology Bryan Pfaffenberger would claim, was anything but revolutionary.[25]

A sweeping transformation in the global political economy of the 1980s will take place, and not by chance but ideology, with the rise of commercial software and personal computer businesses. Shaped as much as it was by intellectual property (IP) with rapidly expanding enclosures of "immaterial goods," this was also a period of consolidation of international computer hobbyist scenes that had branched out from their radio amateur and phone phreaking ancestors. As IP became the new currency in the corporate game, computer businesses started to have more frequent and unpleasant encounters with hackers—soon to be recast as "computer criminals" with an unstoppable if not "destructive" drive to know more and more about digital technologies. For the emergent industry, hackers represented an inestimable source of expert labor and, simultaneously, an antagonistic force to be combated for their practices of software sharing, social engineering, and systems penetration. Meanwhile, a well-sheltered transnational community continued to carry forward a rather elite circuitry of technical exchange, fostering a software commons by circulating code for the collaborative development of the Unix operating system. An up-and-coming generation of hackers were cutting their teeth at the console by exploring security flaws and internal secrets of various commercial "flavors" of the system.

The ancestral hacker behind the first implementation of Unix, Ken Thompson, would tellingly dedicate his prestigious Turing Award Lecture (1983) for the Association for Computing Machinery to the rising problem of "trust" in computing. After demonstrating instruction by instruction how to deceitfully introduce Trojan horses at various levels of computation—that is, from "userland" applications all the way down to the microcode that runs in a processor—Thompson addressed the media-saturated activity of teenage hackers. "There is an explosive situation brewing," he warned. "On the one hand, the press, television, and movies make heros [sic] of vandals by calling them whiz kids. On the other hand, the acts performed by these kids will soon be punishable by years in prison."[26] It was as if two major genealogies of hacking ran neatly in parallel up until this point, where they crossed paths, colliding with quite distinctive practices and understandings.[27]

The shadowy figure of the "dark-side hacker" would take strong hold of the cultural imagination by the mid-1980s, engulfing the computer common with mystical clout. One of the earliest collectives to gain media attention, the same Thompson originally condemned, was the 414s, a teenage group credited for propelling the passage of the Computer Fraud and Abuse Act (1986) in the United States due to (mostly) harmless visits to computer networks with lax but standard security at that time. In fact, a myriad of hacker collectives from various countries, such as Germany,

Netherlands, England, and Australia, were quite active in this period. Clustered in places where access to personal computers and modems was extended to the middle classes, these collectives were quickly stigmatized as "rebellious youth," organized in gangs to wreak havoc behind the keyboard—a very popular but partial and misguided representation.

In the mid-1980s, tech journalist Steven Levy published *Hackers: Heroes of the Computer Revolution* as a corrective to widespread vilification of the community. In his modern classic, Levy narrates the earliest and most compelling stories of hardware and software hacking at MIT, the Homebrew Computer Club in the San Francisco Bay Area, and the very first steps of the computer gaming industry. The book circulated rapidly among hackers, created instant fans, bridged several generations, and gained international acclaim through numerous translations in subsequent decades. One of its lasting contributions, however, was Levy's elaboration on the "hacker ethic" as a technopolitical ethos that was not codified but could be abstracted from the practical experience of early hackers. According to the tech journalist, core tenets characterized hackerdom, all of which were either based on the computing commons or represented an overt commitment to maintaining one. The first and most important was identified as the "hands-on imperative," with the idea that "access to computers or anything that might teach you something about the way the world works should be unlimited and total." From this cardinal orientation followed all the others—that "all information should be free"; that "authority" and "centralization" should be mistrusted; and that "art and beauty" were digitally possible, because computers "can change [lives] for the better." To date, one of the most inspirational and yet most controversial tenets is the one that concerns socioeconomic and cultural belonging: "Hackers should be judged by their hacking and not," Levy concludes, "bogus criteria such as degrees, age, race, or position."[28] As early hacker circuits were rerouted with a wide range of technopolitical orientations in subsequent decades, the "hacker ethic" would be problematized, revisited, and rewritten across contexts. There was no unified ethic but rather a set of context-sensitive "genres" and moral economies of hacking with distinctive commitments within quite specific contexts of invention—as aptly demonstrated by anthropologists Gabriella Coleman, Alex Golub, and Christopher Kelty.[29]

After the first half of the 1980s, the moral valence of hacking would tilt toward its identification with devious, addictive, and criminal behaviors. The controversial term *cracker* was put into circulation to draw a distinction between "malicious attackers" and the ancestral hackers that Levy first brought to public attention. Collectives that shared commercial software and video games further popularized the notion by circulating "cracked" versions with copy protections removed.[30] We find in this period droves of young technologists who would become police cases as systems' penetration became widespread. Raids became commonplace, targeting individual hackers in the United States, Germany, the UK, and Australia. By the late 1980s, police cases were promptly converted into veritable thrillers of computer exploitation told and retold by journalists—such as sci-fi author Bruce Sterling in *Hacker Crackdown*—as well as hackers themselves in their memoirs.[31] Like Kevin Mitnick, who served an unfair sentence for being labeled the "dark-side hacker," and Robert Tappan Morris, who unleashed one of the first self-replicating internet viruses (worms) and brought down machines of ARPANET and MILNET, many hackers of a much lower profile made it into a flurry of publications, depicted as "bad actors" with addictive personalities. In retrospect, these were romantic times of network exploration that were met with overbearing laws against "computer fraud," followed by the consolidation of the profitable business of computer security.[32]

By the early 1990s, the situation would change dramatically from the early days of the computer underground, as security consultants Dan Farmer and Wietse Venema suggest:

> This seems to be the popular image of a system cracker. Young, inexperienced, and possessing vast quantities of time to waste, to get into just one more system. However, there is a far more dangerous type of system cracker out there. One who knows the ins and outs of the latest security auditing and cracking tools, who can modify them for specific attacks, and who can write his/her own programs. One who not only reads about the latest security holes, but also personally discovers bugs and vulnerabilities. A deadly creature that can both strike poisonously and hide its tracks without a whisper or hint of a trail. The übercracker is here.[33]

Expressions of computer-aided superpowers of "übercrackers" have appeared in the public discourse ever since. Security, cryptographer Bruce Schneier reminds us, is often a fear sell. What is curious to note, above and beyond the media circus, is how the magical power of hackers—here understood anthropologically as a set of persuasive techniques that create a (symbolic) separation between magicians and their audiences[34]—has been widely, yet covertly utilized. We find strong evidence in the do-it-yourself (DIY) media of the "hacker underground": the self-organization of a commons around bulletin board systems (BBS) where hackers not only were socialized, but also where they shared studies and findings with other magicians. It was through the participation in "board culture" that the underground would thrive through commoning that was considered, due to the mad race for software commercialization, a serious threat for IP-based business models.

On the flipside of this rather exclusive collective experience, electronic networks also had many harmful aspects, such as the idiotic (in the etymological senses of both "unskilled, rude, simple" and "private, unprofessional, ordinary") celebration of text files on how to prepare explosives and other deadly weapons, known as "anarchy files," as well as vast streams of misogynistic and racist materials.[35] As Charlton McIlwain reports from his study of the experience of Black Americans in online spaces: "BBS, Usenet, The Internet. Yes, they were creating a whole new world. But it wasn't a question about if and when racism would rear its ugly head in this new world. Racism, fueled by anti-blackness, was already there when it began."[36] It should be well understood that in the minor histories of computing the common is no panacea, but rather an opening to computing otherwise that deserves our attention.

Commoning as Method

My own point of departure in research was an origin myth ("there once was a time when software was freely shared among hackers . . .") crafted with a mastery of persuasive magic to interpellate the disquieted hearts and minds of computer aficionados in Porto Alegre, a peripheral town of a peripheral country. Digital technologies were hard to come by, so everything we could put our hands on, we shared. Computer parts circulated as much as games, applications, manuals, technical magazines, and books. Very few of us had access to computer machines throughout the 1980s, and by the late 1990s we were still gathering in a friend's house, taking turns at the keyboard, browsing the internet for the first time. Running a BBS was a luxury that only the richest had the opportunity to experience: phone lines were traded as valuable property before the first wave of large-scale privatization in Brazil in the mid-1990s.

During the terminal phase of the military regime in the mid-1980s, a market reserve was implemented for computers using the Política Nacional de Informática (National Informatics Policy), and we witnessed the growth of a local industry with a vibrant home-brew scene. Social studies of computing of this period have emphasized how much it all depended upon *cannibalizing* computer architectures from abroad (in this case, through reverse engineering, according to computer science scholar Ivan da Costa Marques[37]). However, the most original experiments included creating our own flavor of Unix, called SOX, from scratch and assembling our first digital computers based on minicomputer architectures of the 1970s, such as the Patinho Feio—which made it into the archive through the work of historian Márcia de Oliveira Cardoso.[38] The sad part of this minor history is that it has mostly disappeared into the oblivion of perishing feed-paper printouts and design drawings in the basements of the first computer engineers.

By the late 1990s a new convivial circuit began to be routed through software sharing—known in our circles as Software Livre[39]—with technical collectives that were formed through the unlikely encounter between social movements and software engineers, government staff, university researchers, and enthusiasts of all ages (though not all genders and socioeconomic classes) to rethink what digital commoning could be from a peripheral standpoint. We started to self-organize computer users' groups to share what we learned through whatever means we had—say, a faded Xerox copy of a copy of a Unix manual.[40] Our nascent community imagined itself through unlikely connections that were forged with the rise of the newest Left in Porto Alegre at the World Social Forum.

Immersed in the technopolitical effervescence of the early 2000s, we found each other through but also against corporate computational orders in pursuit of collective exploration and self-training by downloading at the lowest speeds, testing and studying, or writing and sharing software to the best of our knowledge and infrastructural capabilities. We were certainly not afraid of disassembling technical objects for the purposes of learning. Little did we know that we were embarking upon a long decade of technical activism, articulating political debates and struggles for "digital sovereignty" in all-out opposition to the transnational IP regime. Our genealogy was not that of the Northern California libertarian culture—far from it. We were much closer to the project of political self-determination that we identified in the Zapatistas' reappropriation of communication technologies in Chiapas, Mexico. "Hacktivism" became a key symbol that stood for the solidarity between hackers and social movements with powerful examples of direct action coming from groups like the Electronic Disturbance Theater and the Independent Media Center—set against the socioeconomic evils and upheavals of corporate globalization. The discourse on software freedom interpellated us, but different understandings of "freedom" were at stake in this historical process. The musician, technologist, and main agitator of the Quilombola cultural center Casa Tainã, Antônio Carlos TC, saw the urgency of this digital project early on. "Free!" he once interjected. "For those of us who were enslaved, the question of freedom gets immediately our full attention." And so we persisted *(under)commoning*[41] beyond the hegemonic circuits of computing in the Euro-American world.

Alterglobalized circuits of software sharing have one of their starting points, curiously enough, in a conventional research laboratory under rather ordinary circumstances in the Global North. By the mid-1980s, the project of Free Software (conceived simultaneously as a software development practice, a set of legal devices for circumventing IP restrictions, and an international technopolitical project) came into existence through the common experience of a "moral (and technical) breakdown"—that is, a breakdown in our naturalized, intentional (but limited) relationship with the world.[42] As far as the origin myth goes, the project started at the MIT Artificial Intelligence (AI) Lab with a malfunctioning printer. We are told that a prolific operating system programmer, Richard Stallman, wanted to use his engineering prowess to fix the issue but was prevented by having his access to the code that controlled the printer denied. As we learned from Levy's depiction of the "hacker ethic," denial of access went against a common practice of information sharing among peers.

The annoyance with a failed network printer gave Stallman the opportunity to realize that the "original affluent society" of software sharing—to borrow anthropologist Marshall Sahlins's felicitous expression—was rapidly changing around him. From a minor technical disruption, Stallman diverted his attention to one of a more serious and profound moral nature: having his access denied represented a major interruption of the common circuit to which he belonged. From his vantage, the denial represented a form of betrayal. "To be a hacker at the AI Lab meant that your ethical code was driven by progress of the computer code—it was wrong and almost evil to keep code and computational resources for yourself. Hackers respected one another because they were good at what they did, not because they had titles and money," he would protest alongside another ancestral hacker, Larry Wall. "This led to profound conflicts with other ethical systems, particularly those who gave supremacy to individual ownership of ideas. From the hacker perspective, to keep an idea or a new program to yourself was the same as a spit in the eye of everyone else."[43]

Common Circuits is a product of these experiments with commoning across unequal contexts, which I partook in firsthand at the intersection between computing and anthropology through numerous border crossings. After living through and studying technopolitical cultures of Software Livre in Brazil, I found myself working closely with "cyborg anthropologists" in California to examine moral economies of science and technology.[44] At the crossroads of distinct technopolitical histories, I came to inhabit a position between cyborg and cannibal traditions. Cannibal intellectuals in Brazil had historically an attitude of non-submission but also, in the words of literary critic Haroldo de Campos, of creative expropriation, recontextualization, and transformation of the European canon.[45] I realized over time that being tactically placed in between these traditions gave me the opportunity to study instrumental forms of economic rationality while participating in emergent forms of convivial relationality. I found in convivial projects an interruption, albeit temporary and localized, of the radical monopoly of "Big Computing," thus offering a point of departure from the abstract utilitarianism of computing (as a discipline) to the practice of hacking in self-organized collectives (as a generative indiscipline).[46] At this crossroads, I set out to examine practices of commoning through active participation in the open technology circuits that I document in this book.

The interface between cyborg and cannibal anthropologies helped me realize a broader potential for the discipline through their persistent indisciplines: fostering collaborative ties for conceptual work across disciplinary borders, staking the importance of situatedness with the recognition of partiality in our knowledge practices, and opening up the field for multi-locale research of emergent technoscientific domains of practice. What cyborg and cannibal anthropologists investigated through renewed ethnographic engagement was our shared sociotechnical condition that could not be included in the disciplinary agenda (an anthropology without the "primitive") until more recently. Informed by this line of work, I dedicated myself to studying the transposition of the economy of software sharing to the domain of hardware design and fabrication, a very unlikely process, we must say, taking place in community spaces for computing worldwide. This problem led me to study the passage of a moral economy of software to hardware commoning under the guise of Open Hardware.

The more I learned about contemporary, self-organized convivial spaces for computing, the more evident it became that active participation in technical exchange was a fundamental condition for the study of commoning. For the purposes of multi-locale ethnography, I followed technical and political activities in situ at hackerspaces, as well as in motion as I traveled to conferences as a participant and, at times, as an organizer, between 2013 and 2017. During preliminary field research, I identified the Pacific Rim as an area of key importance for concentrating closed circuits of migration of IT professionals and exchange of open technologies. The hackerspace network in the Pacific Rim embodies a technopolitical location of concentrated Western investment and projection, helping to reconfigure geopolitical dynamics. To study community spaces through their placemaking practices, I drew from an understanding of ethnography as "translation circuitry"—that is, "a circuit whose input is the ethnographic, multi-locale experience and whose output is an ethnographic representation that is constituted by symbolic links to other forms of representation."[47]

The criteria I applied for selecting community spaces in the Pacific Rim had to do with the importance of their projects vis-à-vis the broader hackerspace network. Not every community space participated in the social life of the network; each location created specific conditions for sharing wildly different sociotechnical experiments. There were substantial contrasts in terms of how influence and prestige were distributed across locations, projects, and members. Differences, more often than not, masked underlying frictions of class, ethnic, and gender identifications and belongings. Whereas many spaces represent regular hobbyist clubs without much of an active connection to broader transnational circuits of exchange, expressing a much longer genealogy that branched from the first ham radio to present-day robotics clubs, the ones I describe here were characterized by a strong influx of open projects and cosmopolitan technologists that became known well beyond their communities of origin.

In preparing this book, I drew primarily from projects and trajectories of active community members in San Francisco, Shenzhen, and Tokyo, in addition to several local and international hacker conferences. In what follows I will explain why placemaking matters for the contemporary reinvention of hacking as a cosmopolitan practice of commoning.

Common Spaces and Places

"A hackerspace," I was told by one of the founders of Tokyo Hacker Space, "relies on three variables: its space, its people, and its projects." From its people, hackerspaces become sites for socialization and self-directed learning with public performances of technical knowledge. Through the initiative of its members, a hackerspace expands beyond geopolitical boundaries, hosting regular gatherings in bustling urban centers. Through member projects in circulation, a hackerspace travels far and wide through tangible and intangible media, articulated in instances of virtual and actual exchange. Based on its physical space and equipment in place, a hackerspace affords certain projects while constraining many others, serving as a magnet as much as a centrifugal force for the uninitiated. Some places are well-stocked with donations from the most powerful tech companies on the planet, while others are maintained mostly from the means of their members at the periphery of peripheral circuits of computing. Through their spatial and technical affordances hackerspaces constitute nevertheless simultaneously imagined spaces and convivial places. They are conceived by their members as heterotopic spaces for reclaiming information technology and sharing technical expertise—which is to say, they are spaces of difference in the context of corporate circuits for computing in educational and professional spheres.

One of the origin stories of the global hackerspace network starts with the publication of a blueprint for new computer clubs: the "hackerspace design patterns" delivered by two German hackers, Jens Ohlig and Lars Weiler, at the influential hacker gathering Chaos Communication Congress (24C3) in 2007. The hackerspace design patterns provided the general orientation not only for organizing places of conviviality but more importantly for avoiding conventional separations between designers, builders, and users. The productive capacity of a community space is perceived by network participants to have central importance in creating social ties: hackerspaces that are not active are deemed uninteresting and undeserving of the title. Through the circulation of (privileged) people and (open) projects, hackerspaces take on a life of their own. First encounters often involve asking what someone is working on in order to create an opportunity for demonstration of a technique or technology of shared interest. The creation of a "pattern language" for hackerspaces represents an important contribution for identifying challenges and proposing infrastructural solutions for facilitating exchange with, for example, infrastructures with "radical openness"; horizontalist conflict resolution; independence from governments and companies through membership-based funding and donations; sustainability through active volunteering; "creative chaos" for enticing practices of hacking as expressions of poiesis; and distributed responsibility for the governance of the space through consensus-based decision-making. The hackerspace-design-patterns catalog included key issues of concern in the creation and maintenance of new spaces ("sustainability patterns"); autonomy from universities, companies, landlords, and neighbors who may not be happy with the space ("independence patterns"); sustenance of active interest and participation as well as renewal of the membership body ("regularity patterns"); governance and interpersonal conflict issues ("conflict resolution patterns"); and finally, the basic ingredients for keeping the members running wild with their projects, such as a good supply of mate-infused[48] soda drinks and materials to hack on ("creative chaos patterns").

The orientation for creating and sustaining spaces through "design patterns" has been widely debated and has given rise to important derivations, such as "feminist hackerspace design patterns," drafted after numerous events of harassment against gender minorities in the network. In her study of community spaces, communication scholar Sophie Toupin analyzed how openness, one of underlying principles of the design patterns, was problematized through the creation of feminist hackerspaces.[49] As an anthropological intuition would have us suspect, the original design patterns were key in instituting the imagined global network, but their application was anything but straightforward, given insurmountable differences in political histories and available expertise and infrastructure, in addition to events of discrimination that ignited multiple forms of friction and conflict. Over time, hackerspace members were faced with the hard lesson that commoning is much easier said than done in the context of rampant individualism, sexism, dispossession, and socioeconomic precarity that intervenes in communitarian experiments worldwide. The sociotechnical foundations upon which common spaces were built was highly disparate from the start. Community spaces in global cities, for example, were much better off than others in their capacity to sustain a fragile common, based on the relative socioeconomic privilege of their members. The distribution of computing skills was highly unequal as well, mirroring divides of gender, race, ethnicity, and socioeconomic class. Despite these

. profound challenges, the design patterns served as a generative guide for implementing learning webs in community-driven, self-organized educational spaces.

The importance of studying and contributing ethnographically to the informal, marginal, and self-organized spaces for computing has to do with the fact that they are responsible for the spatialization of hacking. This is an emergent process that deserves careful consideration as commoning is practiced in more perennial spaces—not to be identified, we must say, with the process that political theorist Wendy Brown has identified as the "self-cannibalizing tendencies" of the so-called sharing economy that involves, in fact, very little sharing but aggressive rent-seeking that drives mass dispossession from Silicon Valley outward.[50] To demonstrate this crucial distinction, we will start our exploration of the hard challenges of commoning in the highly exclusionary, high-tech hub of the San Francisco Bay Area. We will examine the social experiment of radical openness and horizontalist governance at Noisebridge, a hackerspace that is globally taken not only as an inspiration for community-building but also for its community-unraveling contradictions (chapter 1). From its extensions through the circulation of hackers and their projects, we will follow the transnational circuits that inform central sites of digital production in Southern China. We will turn to the experience of Chaihuo and the hacker group Shenzhen Do-It-Yourself (SZDIY) to discuss how convivial circuits can interrupt entrepreneurial pressures in highly unequal contexts (chapter 3). We will see how this community space was made relevant to the global network for creating a space for self-organized local collectives through cosmopolitan linkages with itinerant hackers and Open Hardware businesses to convert gifts into commodities. Finally, we will follow the unlikely response of Tokyo Hacker Space (THS) to the Fukushima disaster (chapter 5). An otherwise ordinary computer hobbyist club, THS rapidly became a critical site for the circulation of open technologies and technologists. At the onset of the nuclear disaster, THS became a laboratory for serious bricolage of open radiation-monitoring devices, thus serving as an influential example of the technopolitical alternatives that thrive in autonomous spaces.

Based on the ethnographic engagement with these three community spaces, I caution readers against the recurrent error of equating entrepreneurial apples with hackerspace oranges—which is to say, against misreading our examples as entrepreneurial circuits of "digital innovation." An important and frequently asked question concerns the classification of any particular space: is it a hackerspace, makerspace, or a fablab? Within the global network, each space I describe draws from local genealogies in political and computing histories as they intersect with transnational circuits of open technologies. Variants of "hacking" and "making" can be as culturally and technically distant as a politicized, autonomist convivial zone or a coworking space for "flexible" and precarized IT workers; a space to match computer programmers with investors to create start-up ventures; or merely a shared place to meet, functioning as a hub for expatriate computer aficionados in global cities. Hackerspaces are particularly important in this debate because they are characterized by a form of indiscipline that consists in building social and technical ties for exploring digital alternatives. They are not innovative in the industrial sense of the term, but they can, at times, be anti-innovation.[51] They often interface with entrepreneurial circuits and give rise to companies and services of all sizes, but they are not built for market purposes. This is an important distinction that is scantly observed in the existing literature. The hackerspace network is responsible for *spatializing* the practice of hacking in distinctive ways, but also for shaping conditions for technologists within and beyond the Euro-American axis to be cultivated as hackers. And this is the subject of another core component of this book: the *personification* of computing expertise in spaces for computing otherwise.

Personhood through Commoning

If it is between the public and the private spheres that we identify the common, it is between the personal and the technical that we find the moral circuitry through which technologists are formed. When it comes to practices of commoning that extrapolate market logics, we are reminded by Marcel Mauss, founding figure of economic anthropology, that it all comes down to mixtures: persons and objects, persons-in-things, and things-as-tokens-for-persons create and sever social ties, create and re-create sociotechnical arrangements that escape utilitarian justifications for why people engage in exchange practices. Persons and things come "out of their spheres," as Mauss poetically suggests, "to mingle." And this is "precisely the exchange and the contract" that creates a social form distinct from established institutions of computerdom. The recipro-cal dynamic that makes this possible is that of technical objects as gifts, circulating to create pathways ("traces") linking hackerspaces.

In the common circuitry of the hackerspace network, technical objects are made and remade, but so too are persons through their trajectories of circulation and their practices of exchange. Technical personae are cul-tivated in specific political histories, made through particular sociotech-nical trajectories, individuated in relational arrangements. In the debate over what makes technologists hackers, it is my contention that common-ing is foundational for self-cultivation in and through computing—one of the most important means for educating one's attention through active participation[52]—thus serving as an interpretative key for examining the formation of technical personhood.[53] It is in this particular sense that en-gaging personal trajectories can help us illuminate intricate technical and experiential dimensions of hacking. It is through the study of commoning that we get to understand how technologists are cultivated otherwise.

Under the guise of moral demands for cultivation in spaces for community-oriented learning, little is currently known about the idioms of ethical reasoning[54] of computer experts beyond the liberal tradition deftly described by anthropologist Gabriella Coleman.[55] Despite the exis-tence of a solid body of literature on the historical foundations of hacking in the Euro-American context,[56] substantial work remains to be done on the cosmopolitan dynamics of the *alterglobalization of computing expertise*, the process responsible for spatializing the practice of hacking with the emergence of expert collectives beyond the Euro-American axis. While these collectives are still entangled in many ways with their counter-parts in the hegemonic centers of digital development, their trajectories are often marked by contrasting and unequal conditions for cultivation of (noncredentialed) expertise. As such, their technopolitical histories describe a different orientation toward questions of openness, autonomy, and conviviality. It is in these distinctive contexts that computing exper-tise is cultivated otherwise.

Common Circuits describes personal narratives of technical cultivation to give us access to the ways in which hackers reflect on their moral expe-riences. By retracing the trajectories of three influential hackers—Gniibe (Japan), NalaGinrut (China), and Mitch Altman (US)—we get a better sense for how moral and technical orientations are articulated through practices of commoning. Here I invite us to redirect our attention to an important but often forgotten dimension of computing expertise by asking what the moral sources for the practices of commoning are. What binds technologists in moral economies beyond corporate circuits of computing? Each personal trajectory we will trace brings to light the distinctive experiential grounding of hackers as they "learn to learn"[57] through the exchange of open technologies. We describe the constant search for community through which Mitch, one of the founders of Noisebridge, became one of the global spokespersons for the hackerspace network (chapter 2). In the tech pilgrimage site of San Francisco, Mitch promoted community as an antidote to the never-ending gold rush where computing stands for the ultimate symbol of "disruption," leading to dispossession and displacement for those who are not connected to computing and financial industries. His trajectory is one of political cultivation in queer spaces, communes, and happenings from the 1970s to the present. We turn next to the trajectory of NalaGinrut (chapter 4), Chaihuo resident hacker and one of the Shenzhen Do-It-Yourself (SZDIY) founders, who combines the practice of software and hardware sharing with questions of prosperity, familial commitment, and community-making in the cosmopolitan center of Shenzhen. In NalaGinrut's experience, we find the articulation of hacking as a spiritual calling at the crossroads of neo-Calvinism and neo-Confucianism in postsocialist China. His work is perceived by the local community as fundamental for bringing young Chinese engineers to participate in common circuits. Finally, in the technical landscape of Tokyo, we engage with the trajectory of Gniibe (chapter 6), a veteran hacker whose software and hardware development practices are interpreted curiously through the Golden Rule in its negative form: hacking as a moral imperative that prevents expert technologists, such as himself, from exercising power over others, or as he puts it, "hacking for doing no harm."

As we retrace these three trajectories with their moral sources for ethical reasoning and self-cultivation, we position ourselves to better identify differential conditions of possibility for commoning. In these three cases, hacker personhood is made body and skill[58] through contrastive spiritual orientations, ethical struggles, and political relationalities as much as by their common sense of intimacy with computer architectures, programming languages, network protocols, and digital tools. Through active making and mixing of technical persons and objects, the common opens up as an interstitial domain of exchange in the midst of institutional and corporate circuits. Narrative articulations of what it means to be a hacker represent the output of the (context-dependent) transformative circuits we abstracted from individuated trajectories. From the joint exploration of the spatial and personal dimensions of hacking, we find the last process that we will examine in this book: the formation of a technopolitics where hacking stands for the practice of *hacking* ties.

Technopolitics Otherwise?

As the 1980s came to an ending with Cold War–era "closed comput-ing,"[59] a watershed event, the Galactic Hacker Party—also known as the International Conference on the Alternative Use of Technology (ICATA '89)—brought together hackers from various generations, nationalities, and political orientations for the first time.[60] In ecstatic anticipation of the gathering, the Dutch techno-anarchist collective, Hack-Tic, disseminated a call for participation in their zine of the same name. "During the summer of 1989 the world as we know it will go into overload," the organizers promised. "An interstellar particle stream of hackers, phone phreaks, ra-dioactivists, and assorted technological subversives," coming from most corners of hackerdom up to that point, "will be fusing their energies into a media melt-down as the global village plugs into Amsterdam for three electrifying days of information interchange and electronic capers."

And so they did. Over one thousand participants formed a *communitas* of technical exchange accompanied by debates of pressing political mat-ters. Not only the scale was unusual for hacker conferences at the time, but so were the internationalist angle and political sharpness of the debates. Telephone uplinks were established for hackers, activists, and academics interested in examining the political use of information technologies in capital cities, such as Nairobi, Moscow, and Wellington. Computer net-works were wired in a public hacklab for the exploratory delight of the con-ference participants. Ancestral figures of the Homebrew Computer Club, John Draper and Lee Felsenstein, were guests of honor and took the stage to share their experiences in phone phreaking and community-driven net-works. On the table was the potential of digital technologies to be used in democratic processes, but also for mass surveillance, long before we got to witness the demise of cyber-utopianism on both counts. Ecological activ-ists, by their turn, planted the seeds of an urgent debate that would mark future problematizations of the digital: the double bind in which many activists find themselves when defending the use of computer networks for mitigating environmental impact while contributing to environmental pollution through the consumption and disposal of computers and peripherals. Particularly fitting for the hacker party was Terry Gilliam's film *Brazil*, whose screening was meant to serve as a warning, the co-organizers remarked, about a potential future with "too many machines and too little humanity."[61] Fascist software ("Naziware") was controversially brought to the stage, examined, strongly objected to, and promptly rejected. The contemporary anarchist classic *bolo'bolo* (1983) in its Dutch translation was presented as well, but as a miraculous antidote to fortify the imagination of alternative futures to capitalist and Soviet modernities. The publication sold out amazingly fast as excerpts were read out aloud over the PA system. And a historic heated panel was held on the ethics of information disclosure involving a co-founder of Chaos Computer Club (CCC), Wau Holland, and Pengo, a club member who had been ousted for selling information to the KGB that he and his friends had obtained from foreign computer networks. "Responsible handling of information," Wau explained, "has enabled us to use it very effectively in a political sense." He concluded with a reminder about their mission: "I think of the CCC as a *public* service instead of a *secret* service," drawing an important distinction that would inspire scores of future hacker activists. At the end of the conference, the ICATA '89 declaration was drafted to address the "deeply [disturbing] prospects of an information technology let loose by economic and political actors without democratic control and effective popular participation." The declaration included a telling contribution from University of Nairobi: "The current discussions on alternative uses of computer technology are guided by the same misleading assumptions that dominated the modernisation theories of the 1950s—particularly the notion of information technology as an independent variable." After all that had been presented and discussed at the conference, how could computing be defined as an independent variable? The declaration proceeded with a prescient warning: "Computer technology is a dependent variable—its effectiveness will largely be determined by the existing social conditions. As a result, more computers means more global inequality."[62]

We started our pathfinding in this book with the hint that we could conceive of hacking as a contentious variable with at least six decades of experimental history. If we were to take into ethnographic account the spatial and personal aspects of the practice, I suggested, we would be able to expand considerably our understanding of who counts as a hacker and what counts as a hack. From the postwar period to the present, we saw how computing was practiced and transformed into different objects of discourse, indexing different infrastructures, technopolitical practices, and moral experiences. The valuation of hacking shifted between the defense of the virtuous practice of computer exploration and the criminalization of the underground study of information security. Little did we know that, at the turn of the century, hacker tools and techniques would be vested in digital infrastructures of planetary scope: from (open) cryptographic technologies for secure communications to basic internet services with (mostly open) system software. In the process, hacking would shape-shift once again to become strangely profitable and desirable for markets and government contracts: amassing online vulnerability databases, creating independent mailing lists for sharing expert information (outside and beyond credentialed expert circuits), and bundling infosec tools in software distributions for the study and exploitation of computer networks.[63] Barriers to entry into hackerdom were brought to the lowest point at the same time that, paradoxically, sociotechnical systems increased in complexity. The hacker underground was professionalized and contested, creating room in the spotlight for reactionary and rebel assets to appear: the oppositional figures of the state-sponsored hacker and the politicized hacktivist.[64] Free and Open Source projects succeeded in popularizing their collaborative development tools and methodologies, becoming less combative as they provided their not-so-secret sauce to nascent internet companies, empowering infrastructures and research and development teams for the anti-common of Silicon Valley. The fierce opposition to "corporate lock-ins" of the early Free Software activists slowly gave in to the creation of what philosopher Pierre Dardot and sociologist Christian Laval called the "ersatz commons" or the "quasi-commons" of corporate Open Source projects.[65] Between the booms and busts of the computer industry with rapidly expanding sociopolitical and ecological implications, "hacking" became a critical practice with indisputable urgency, for it allows us to see in clearer relief major interplays between the moral and the technical, the personal and the computational, the institutional and the convivial. Hacking became, in sum, central to our understanding of contemporary technopolitics.

In the upcoming chapters, we will retrace the common circuit of hackerspaces through what I call the *alterglobalization of computing expertise*, a scale-making process that re-creates the practice of computing within and beyond the Euro-American context. In the cosmopolitan trajectory of community spaces and projects, we find an important rearticulation of hacking that contrasts with well-established popular definitions of the practice as shadowy computer crime. To the standard economist's shock, we encounter a worldwide range of practices that place in common technical objects, services, and infrastructures of high market value, where hacking consists of an attitude with an openness that stands for, as philosopher of technology Gilbert Simondon notes, "respect for the work of the others."[66] By virtue of a hack, in this sense, we obtain backdoor access to a domain of moral orientations concerning shared technical objects under the charged symbols of "collaboration," "autonomy," and "openness." More than a symbolic act, we will see how hacking generates sociotechnical ties through situated practices of commoning, performing an overlooked binding function that creates and sustains and also dissolves technical, moral, and political ties. Hacking establishes, in other words, technopolitical connections of transnational scope among cosmopolitan technologists in community spaces but also severs ties with entities that are perceived to be adversarial to the purposes of collective exploration. Depending as much on its moral as on its digital circuits, open technical objects circulate widely to find collaborators, embodying invitations for practical engagement. Hacking *ties* technologists and technical objects, in sum, but it also serves to *hack* ties of institutional control to create and sustain common circuits. Ultimately, it is the pursuit of alternative technological futures that organizes the technopolitical formation we examine here.

In what follows, we will probe the noisy signals of *spatialization*, *personification*, and *politicization* of computing expertise to account for the hard challenges of sustaining the common that community spaces have struggled to create at global and yet partial scales.

ONE

Noisebridge

An Experiment in Radical Openness

Technologists are often interpellated in Silicon Valley with a moral question: "Are you a zero or a one?" At Noisebridge, computer technologists are neither zeros nor ones. They are situated in the interstitial space between political collectives and open technology projects, carving out a space of relative and temporary autonomy and linking their organizational practices to a long-lasting genealogy that literary scholar Jesse Cohn describes as the anarchist culture of resistance[1] through convivial placemaking and independent media-making. Their technopolitical experiences are, in short, non-computable.

Noisebridge opened its doors in the historic working-class Latinx neighborhood of San Francisco's Mission District in 2008. As one of its founders recalls, its purpose was not only to create a space for locals but also to start a "hacker embassy" where members of the international community could gather. The hackerspace was built upon the socioeconomic privilege, material prosperity, and countercultural legacy of Northern California.[2] Supported almost entirely on membership dues and small donations, the space was bootstrapped by volunteers who built the workbenches, gathered an impressive library, set up servers and equipment for

26

microelectronics work, equipped a workshop for metal and woodwork, and remodeled a rental property by reworking the floor, repairing the electrical installations, and wiring networks—a collaborative effort well aligned and not at all unusual for community spaces elsewhere. After growing out of its first space, Noisebridge secured in 2009 another nearby rental property with twelve-foot ceilings and 5,600 square feet of floor space for $3,600 a month for an initial three-year lease in the heart of the Mission District.

For the Bay Area tech community, it would be fair to say that Noisebridge figured, for better or worse, as a wild card. It was a place with magical magnetism, an impossibility in the context of San Francisco's Big Tech and startup culture, situated in a zone of simmering class antagonism and autonomist politics too intense for many of the "techies." The Mission was increasingly a place of contrast and dispossession in large part due to the steady influx of young professionals from the tech industry, some of whom would come visit Noisebridge and, every once in a while, engage in its social experiments.

In his study of the political economy of the San Francisco Bay Area, geographer Richard Walker examined the mounting inequalities that transformed the city into one of the most unequal in the United States. While rampant evictions disrupted more than 35,000 households in the previous two decades, compounded socioeconomic inequalities are reshaping the city with a widening wealth gap, housing insecurity, and wages that are incompatible with the sharpening cost of living.[3] It was hard to find another neighborhood in the city where the landscape attested as drastically to the process of hyper-gentrification as it did in the Mission. In an accelerated rhythm of racialized dispossession, Latinx and Black families were regularly pushed out to make room for tech workers, managers, and even CEOs of major internet companies. In the eye of this storm, Noisebridge served as an unlikely stage for numerous debates and activities involving the anti-gentrification movement. Meanwhile, expensive boutiques, fancy cafes, and gourmet restaurants got firmly planted in the neighborhood just a few blocks away. Evangelical churches conducting services in Spanish, migrant community support organizations of various stripes, remittance services, pawnshops, and single-occupancy hotels, alongside a wide array of small Latinx-run businesses with street vendors selling popsicles, fresh fruit, copied DVDs, and churros were distinctive parts of the landscape. The Mission was also a political location for collectives such as Food Not

Bombs and Homes Not Jails, as well as events such as the Bay Area Anarchist Book Fair and Hackmeet. They all came together in the space-time of mixed political histories and persuasions that were most definitely not shaped by the capitalist "libertarianism" of Silicon Valley.

For the broader hackerspace network, Noisebridge became known as a place of inspiration but also a counterexample to mainstream computing, a place where one could find people from various socioeconomic strata with different degrees of cultural and technical capital—a diversity not found in other community spaces in the Bay Area. Among members and itinerant nonmembers, we found elite university–educated young adults, startup founders and employees, and well-established engineers all the way to political activists, NGO workers, and artists of different age groups, sometimes unhoused. Despite this wild demographic distribution, members of the local Latinx population were conspicuously absent from the space, with rare exceptions. It was common for marked differences of socioeconomic background and technical expertise to be the sources of tension online and offline. Across marked socioeconomic, technical, and cultural differences, various in-groups would share the space but would not come together to participate in the same circle of exchange, except for lively public events: programming classes, tutorials on digital fabrication, or hardware workshops.

The radical openness of Noisebridge—seven days a week, twenty-four hours a day without closing as long as someone with a key was there—was often celebrated when communicating the importance of its sociotechnical experiment beyond the boundaries of its immediate collective. This form of openness, however, was highly disputed and eventually rethought: in many hackerspaces, when not lauded for being an exceptional place, Noisebridge was disparaged as a place for the "homeless and Arduinos"[4]—that is, for its contingent of the unhoused, but also for neophytes to learn hardware prototyping. Insurmountable tensions led its collective to experiment with various forms of access control over time. Due to its open-door policy with wide distribution of gate codes and physical keys, conflicts often arose around the perception of "overuse" and "abuse" of facilities and tools. It was common for members to complain about stolen donations, tools, and parts. Despite these occurrences, the openness of Noisebridge was otherwise highly generative of spontaneous collaborations and forms of exchange among strangers who would come to the space to work on shared projects.

Standing as a heterotopic space[5] for intersecting art, politics, and computing, Noisebridge aspired to convivial technology. Its landscape was formed by a wealth of electronics and computing equipment, used machines and peripherals close at hand for those with skills or curiosity to use and repurpose them. Its library was impressively well stocked on topics of computer programming, electronics, and hacking. It included copious volumes of feminist literature, intellectual property critique, and science fiction, carefully curated by Liz Henry, an organic member of the Anarchafeminist Hackerhive. The walls of the space were covered with messages, slogans, stickers, and posters with allusions to political issues. To be physically present at the space was to be surrounded by an atmosphere of "creative chaos" and generative indeterminacy that suggested one of the most important design patterns for hackerspaces worldwide. Walls were ornamented with stencils, painted and repainted constantly, just to get recovered with colorful murals celebrating inspirational figures for the community, such as Margaret Hamilton, Nikola Tesla, and Chelsea Manning. Posters of hacker conferences, zines, and pamphlets pointing to anarchist and queer politics could be found scattered around the space. Unlike the vapidity of the Facebook headquarters' decoration with its public relations twist on the notion of "hacking" as a commercial imperative for "moving fast and breaking things," Noisebridge members espoused distinctive versions of what it meant to "hack." The multivalence of hacking quite often made for a lively topic of conversation, as well as for heated arguments. Overall, the landscape was suggestive of its political orientation toward conviviality by way of shared slogans, high availability of shared tools, and openness to peer-to-peer learning in free workshops and classes. Contemporary political issues, as well as autonomist political sensibilities, were present on the walls of the space (figure 1.1).

FIGURE 1.1   Stencils and inscriptions at Noisebridge

Differences could be found around the space with respect to the way activities were spatialized. Right by the entrance was the "hackitorium" (often spelled "hackit-torium") with a set of long tables where people would meet up, talk, or write code. Conversations were generally about technical topics and sometimes about political issues, but they easily turned into battles of wit and bravado in which interlocutors performed the knowledge of a particular technical object. In front of a former dark-room for photographic work and next to the library, where we could find bar stools and a high table, was a place usually occupied by more anonymous or ephemeral users, where people would normally concentrate on their work using headphones, which was not necessarily software coding. The same applied to the back room, Alonzo Church (named after an influential computer scientist), mostly reserved for public classes and talks, and the kitchen, which was occupied by more transient visitors. The server closet holding the network infrastructure was an informally forbidden area for "do-ocratic" experimentation, given its control and oversight by a tight-knit group of Noisebridge system administrators and early members. Spatial arrangements changed very fast and drastically: they were, overall, quite ephemeral.

The first time I visited the space, I was offered a tour by a white, middle-aged man in dreadlocks, the same figure who a year later would be asked by other members (including myself) to leave the space due to accusations of sexism and "disruptive behavior" (for catcalling women in the streets of the Mission). He was identified by other members as an Occupy activist and member of the squatters movement who hung out at Noisebridge on a daily basis. After walking around the place under his guidance, I entered the elevator with a starry-eyed friend who came with me. He then joked while gesturing as if he held a camera: "I am a Fed and I am taking a picture of you!" I responded with a half smile as I tried to fathom what he meant. Eventually, I learned that privacy was a cherished value at Noisebridge. Video and photo cameras were not usually welcomed except for events with remote participation and live video streaming for the popular show-and-tell "5 Minutes of Fame" in which community members stepped up proudly to talk about the (actual or fictional) projects they were working on.

For my initial four weeks of daily attendance, I had no meaningful connections to the space, just a few acquaintances who had helped get Noisebridge going when it was in a small room near the intersection of 16th and

Mission Streets and who had since given up attending because of recurrent problems regarding the open-door policy. I met them through the International Free Software Forum (Fórum Internacional Software Livre; FISL) that I helped to organize for many years, and we kept in touch through various friends we had in common. In my interactions with Noisebridge regulars early on, I remember just a few other instances of being half-jokingly asked before introducing myself if I was "a Fed," when it was much more likely for me to be a Mission local. A common topic of discussion in the space, alongside discussions of political, artistic, and technical matters, was the wrongdoings of the government and its repressive forces. One often overheard discussions about politics, FBI raids, and NSA spying—particularly in connection with the cases of Wikileaks, Chelsea Manning, and Aaron Swartz—and in anticipation of the leaked documents by Edward Snowden and the outcry regarding the massive surveillance apparatus deployed by the United States and other nation-states. These informal conversations interpellated Noisebridge regulars to take a stance on technopolitical issues.

The sociotechnical life of the space drew a powerful but unlikely crowd of artists, system administrators, software developers, political activists, scientists and science students, university dropouts, and curious tinkerers without formal training. The expert and activist constituents of Noisebridge regularly hosted cryptography workshops and dedicated a substantial part of its network infrastructure to anonymous internet use, supported technically and financially by volunteers. Projects with combative dimensions, such as Noisetor (a nickname for the robust local server infrastructure supporting Tor, a network for privacy and anonymity), were native to the space, inspiring technologists in other community spaces as well. This anti-hegemonic approach to computing incarnated in the value of self-hosting of digital infrastructure for anonymity and privacy caught the attention of the FBI, whose agents have paid quite a few visits over the years. I have heard the stories of their recurrent visits, but have never experienced one myself. They often came looking for the "person responsible" for the network, which, of course, was everyone and no one in particular.

Whereas most newcomers who came for free hardware workshops and programming classes left after the activity, a small number would stay and became integral to the political life of the space. I was one of those participants during the period that I lived in San Francisco to study open technology projects. I was keen on learning but also on sharing what I knew as much as I could. Participation intrigued me deeply as I felt how important it was for Noisebridge to thrive as an experiment in commoning. As soon as I started to make more sense of the space and its activities, I decided to engage more actively in the recurrent event Circuit Hacking Monday, helping with what I could by arranging soldering irons, learning to make little metal stands for them, and responding to questions on basic issues regarding soldering and the placement of electronic components. At that time, I could not help as much as I was learning to learn myself[6] (which anthropologist Gregory Bateson called "deutero-learning"[7]) through an immersion in the process of hardware tinkering with seasoned engineers. I basically limited myself to relaying information on how to get certain things done as I had been instructed. There was no shortage of people to help and to get help from during these gatherings of rare effervescence, which led me to understand how essential they were for the pattern of interactions they facilitated. Interactions were mediated by technical objects and organized around what I came to identify as the "pedagogical mode" of exchange. When meeting people, parties to an interaction would take turns as conveyor and recipient of technical information, but the interaction was not limited to what anthropologist Jean Lave has identified as two major but misguided approaches to learning: first, that education is the synonym of "transmission of knowledge," and second, that mental representations lead eventually to the internalization of knowledge.[8] Hackerspace exchanges involved rather a practical orientation on how to get a certain task done, delivered with incredible patience and detail. There was always something to learn from someone or something to convey, whichever came first in the order of interactions that cascaded over time. Exchanges were always grounded in actual practice, even when they were self-directed; they rarely stayed in the realm of "pure" technical forms and ideas. At the end of my research, I realized to my surprise that this particular pedagogical mode served as a frame for the process of learning in other community spaces as well.

Participation in common activities led to further access to experts and their open projects. Based on my previous experiences in open technology workshops, I was eager to learn more about the novelty of Open Hardware through the practice of assembling kits alongside novice and skilled engineers. I was particularly puzzled (and ecstatic) about the possibility of transposing the logics of software exchange to a hardware commons.

It was this puzzlement that initially opened the door for the study of the computing common in its spatial, technical, and moral dimensions.

Before one of the regular Circuit Hacking Mondays, I happened to be waiting by the gate when I was approached by one of the founding members, Mitch Altman, who smilingly asked, "Do you have a key?" I was not expecting this question, so responded I did not. He asked, "Do you want one?" to which I promptly said I did "for sure"! He reached into his pocket and handed a key to me, saying, "There you go," with immense satisfaction. In a sea of ephemeral everyday encounters in the space, this gesture felt like a vow of trust, even though it was not at all exclusive. At that point, sharing keys was very common, and in a TED talk on the international hackerspace network, he even dumped dozens of keys on the stage to make a point about the "radical openness" of Noisebridge. The gesture was special, however, for what it communicated: it was an invitation for participation in the life of the space, a social tie that implied shared responsibility for its present and future. I was hooked.

In time I was also offered an electronic key code based on recognition of what I was willing to contribute to the community. In an ordinary evening of bubbling activity, one of the well-known local hardware hackers, Jake Spaz—an inventor and free radio activist—was wiring a new interphone system with a camera feed coming from the entrance of the building. I was interested in the system and the installation process, so I followed it closely but silently, helping whenever asked. Jake needed an Ethernet cable, so he handed me a box of tools to prepare it, which I gladly did with skills I had acquired in high school while working pro bono for a small company to learn about networking. He was startled when he looked at the clock and realized he had to leave promptly to catch the last train. He asked me if I could finish wiring the system, and so I did, running the cable across the building alongside an existing bundle of wires all the way to the front door by Mission Street. The next day he did not mention the wiring but approached me to ask if I wanted a "code"—a shared code that could be punched into the decommissioned pay phone at the entrance to open the gate, which I happily accepted.

Workshops for Commoning

There was an unparalleled collective energy in the space at every hardware workshop.[9] It was unlike any other activity; it changed the atmosphere. At Circuit Hacking Mondays, newcomers would pack the hackitorium, technologists of various stripes would gather around shared tables, and unlikely opportunities for learning would readily open up. Parents wanted their kids to have a piece of the action in the free soldering lessons; artists came searching for engineers to partner with; and engineers with different skill sets offered to help. Circuit Hacking events were special in their conviviality. Their organization, interestingly enough, involved exchanging and intertwining gifts and commodities.

There was a structure to the event that would repeat itself with ritual regularity. Before the arrival of participants, workshop organizers would set up the table with soldering irons and wet sponges to clean soldering-iron tips and put out wire clippers that would be used after the soldering jobs were done to cut excess wire off through-hole electronic components. The instructor was often Mitch Altman, but sometimes someone else with long experience. The workshop would start with an explanation of the electronics kits for sale. On display were a variety of kits that were simple to assemble in a couple of hours, such as gadgets with blinking LEDs, electronic games, and mischievous projects such as TV-B-Gone, a universal remote control to turn off TVs in public spaces. All the kits were based on open technologies and meant for educational purposes. The instructor would announce the prices and ask which kits people would be interested in purchasing. While people made their choice, the instructor would emphasize that the lessons were always free of charge, but there had to be a charge for the kits to reimburse the inventors for the kit materials, as well as to replace worn-out workshop tools. Some instructors charged a small extra amount to supplement their income, helping them to make a living as independent Open Hardware inventors. At this very moment in the workshop, the gifts of open technology would change register to become products, but that was not the end of the exchange; that was the opening of yet another register of commoning.

After participants had taken seats at soldering stations and had their kits at hand, the instructor would demonstrate in a few minutes how to make a solder joint. The instructor would take a test printed circuit board

(PCB), place an electronic component on it, and proceed to join the component to the board with a small amount of solder, forming a Hershey's-Kiss-shaped joint. Though the fumes were sometimes troublesome at first, there was something incredibly satisfying in watching the solder flow. While the demonstration was underway, safety tips were always imparted, following a change in perceptual framing that anthropologists call a "phenomenological modification." "Wash your hands very well after soldering. Lead is poisonous and collects in people's brains, making them sick and aggressive," the instructor would say. "If people become aggressive, they start to lose their friends, and you do not want to lose your friends!" The instructor would then proceed with more practical advice on how to hold the soldering iron to avoid burning oneself and others. Workshop participants would see how to hold the iron in a particular way: grabbing it firmly as one would a pencil. Avoidance of the metal parts was strictly necessary for obvious reasons, so various moments of perceptual modification would succeed one after another. After holding the iron, participants would have to attend to the heating element, the metal surface to avoid at all times. "If it smells like chicken, you are holding it wrong." Participants were also always encouraged to "avoid breathing the fumes" from the soldering flux that vaporizes by the heat of the soldering tip. The nature of the fumes from vaporized rosin, not lead, would be explained to reassure workshop participants. This crash course would not take more than five minutes on average.

As people settled by their soldering irons, they would start assembling circuit boards on their own. Each kit contained a printed circuit board, a small microcontroller, infrared LEDs, and a small set of passive components. If someone in the group had a question, it would often be asked out loud and people nearby would answer, freeing the instructor from conducting a one-to-many session. Participants often used their laptops or cellphones to check a tutorial on how to place components, observing their polarity and pin-out. One of the key affordances for practical learning was the ability for newcomers to follow the documentation created by the workshop organizers. The documentation included an explanatory section about the techniques involved. For example, the TV-B-Gone kit, by far one of the most popular for beginners, included the circuit schematic, software code, and an explanation of how the universal remote control worked. The design notes displayed screen captures of an oscilloscope, a diagnostic instrument, where the off code of a particular TV was analyzed in detail. Explanation of technical topics would be made accessible, such as how Pulse Width Modulated (PWM) signals were generated to turn LEDs on and off at required frequencies by different TV manufacturers. The documentation included as much detail as one was willing to follow from the hardware itself at the component level all the way to the details of how the firmware code was implemented. The TV-B-Gone kit documentation also included a walk-through of the data compression technique its authors used to fit more off codes into the device. The accessibility of the documentation was a real delight to newbies and experienced engineers alike; it was one of the key conditions for learning to learn through collective study.

When it comes to data compression, we are told, one must look for structure and repetition. Once these two parameters are sufficiently understood, a more economic and efficient way of compressing data can be devised. Two influential engineers in the Open Hardware community, Mitch Altman and Lady Ada, guided us through the steps of good compression in the design notes of the TV-B-Gone.[10] They told us that off codes are defined by predetermined patterns of infrared light, blinking at specific time intervals, so the trick is to first capture and study them. Once codes are captured from target remote controllers and properly documented, it is necessary to create a data structure with the off codes, but there is a catch. The little microcontroller chip that TV-B-Gone uses has a very limited memory for storing temporary as well as non-temporary data. Our hardware hackers had to identify which codes repeat themselves, so that they could create a "look-up table" that stored only the repeated patterns. In the following example, we have an array of values where every line encodes the time in microseconds (× 100) that the infrared light will be on and off. So, {240, 60} in the first line represents 2.4 and 0.6 microseconds respectively:

```
const struct powercode sonyCode PROGMEM = {
        freq_to_timerval(38400), // 38.4 KHz
        {
                {240, 60},
                {120, 60},
                {60 , 60},
                {120, 60},
                {60 , 60},
```

```
                {120,  60},
                {60 ,  60},
                {60 ,  60},
                {120,  60},
                {60 ,  60},
                {60 ,  60},
                {60 ,  60},
                {60 , 2700},
                {240,  60},
                {120,  60},
                {60 ,  60},
                {120,  60},
                {60 ,  60},
                {120,  60},
                {60 ,  60},
                {60 ,  60},
                {120,  60},
                {60 ,  60},
                {60 ,  60},
                {60 ,  60},
                {60 ,  0} // end of off-code
        }
};
```

We can easily see the repeated patterns by following the lines above, but we do not want to repeat them in the code. We want to avoid taking up more space than needed. So, the next step our engineers-turned-hackers took was to create a reference table with the repeated patterns, organizing everything neatly in the memory of the little chip. Every "line" in the look-up table below works as an index (say, lines 0, 1, 2, and 3), which can be referred to in order to reconstruct the long code with all its repetitions:

```
const uint16_t code_na000Times[] PROGMEM = {
        60 , 60,   // line[0]
        60 , 2700, // line[1]
        120, 60,   // line[2]
        240, 60,   // line[3]
};
```

After a few extra wizard steps to convert the codes into binary to save as many bits as possible, the end result is a fully functional universal remote control that reduces the number of bytes of an off code from 107 to 27. With compression, saving space, in practical terms, the TV-B-Gone stores 139 codes for European brands and 136 for US ones (whereas TV sets from the "majority world" get subsumed under these two imperial, catch-all categories of manufacturers). These codes will turn off most TV sets in public spaces, airports, and sports bars—with just a few bytes, relatively speaking. This is a substantial contribution to people's well-being with a "neat hack" that becomes simple—as it happens with the most famous hacks, such as the phreaking discovery of the 2600Hz signal—only if we have someone to explain it to us.

Circuit Hacking Monday workshops were so popular that they ended up being replicated across hackerspaces and conferences worldwide. Being able to (informally) learn at the level of practice alongside non-credentialed experts and apprentices created an important disposition: the ability to assemble and disassemble technical objects without hesitancy at the most elementary level with a do-it-together attitude. One learned fundamentally where to intuitively look for (human or nonhuman) support. After extended periods of interaction, hackerspace members felt that they could venture into repair jobs and that they could help digital technology projects or create their own based on the skills they incorporated over time. A great deal of expert information always circulated informally—information that was hardly shared outside engineering classrooms and corporate labs. It was "placed in common," so to speak, regardless of the recipient; there were no forbidden topics for beginners in this mode of exchange. There was, in effect, a little bit of everything for everyone in the workshop. More experienced engineers would discuss their experiences with specific types of hardware and software development frameworks and tools, while newcomers would benefit from an entry point that was unlike anything anyone would experience in a professional or educational institution. There was also ample room to make mistakes in an atmosphere of convivial exploration. Most workshop participants would leave the event with the satisfaction of having completed the assembly of a fully functional kit. From the gift of open technology into a commodity and then from a commodity back into the space of commoning with the re-creation of a gift based on shared information about (open) technical objects—these steps characterized the open-ended nature of the gift in its three fundamental moments first identified by anthropologist Marcel Mauss of *giving, receiving,* and *giving back*.[11]

Imperatives of Excellence and Do-ocracy

Another key experience with commoning at Noisebridge concerned the implementation of a consensus-based form of governance. In his ethnographic study of Bay Area hackerspaces and makerspaces, sociologist Michel Lallement shows how Noisebridge members prefigured a new "grammar of work" through their social experiments in a loosely structured, "adhocratic" organizational form.[12] Grounded in the autonomist tradition, Noisebridge was often described to first-time visitors as a place with one basic rule from which everything else was supposed to derive: be "excellent to each other."

In the daily life of the space, I experienced the beauty and the limits of this demand for excellence, its interpretative use and abuse, and the potentialities but also the disappointment with disputed versions of what it meant. I experienced, in other words, how circuits of commoning were built, as well as how they often broke down. Excellence was instantiated by the "do-ocratic" imperative that stemmed from the sustained hacker value of learning through self-motivated and independent practical engagement—discussed by journalist Steven Levy[13] in terms of the high value of the hands-on approach. "Do-ocracy" emphasizes horizontality in decision-making with an ad hoc orientation: as problems appeared, norms were created and, after a long period of deliberation, consented to instead of being created and enforced a priori.

I came to understand the do-ocratic imperative through practical activities, as well as through observation and reflection upon conflicts I witnessed over time. At the space, the practice of do-ocracy was supposed to be bounded by an understanding of excellence, which was informally and interactively regulated. Like Mitch, many members had previous experiences with autonomist organizations, open technology projects, early online communities in the 1980s and 1990s, and hacker collectives that contributed to the implementation of horizontalist mechanisms of self-governance. This core principle also derived from one of the key orientations of the hackerspace design patterns, the "sudo leadership pattern,"[14] which suggests that if something has to be done, there is no need to ask permission, one must go on, granting oneself permission, unless the act breaks the basic rule of being excellent. This orientation is also part of the Grace Hopper pattern: it is easier to seek forgiveness than permission. I participated in countless instances of amicable interaction and technical exchange but also witnessed numerous (avoidable) instances of blatant violation of "excellence."

Late one evening at the space, a tinkerer in his early twenties who was relatively new was called out for being disrespectful while working on his project in the main room. He was asked not to activate his experiment—a small but loud Tesla coil. Because his intentions were curbed by people working next to him, he responded with verbal aggression. The matter was then taken to the Tuesday meeting, as was common practice in cases of conflict, following the "conflict resolution" in the hackerspace design patterns. At the meeting, the group questioned the young tinkerer about his intentions. One member reminded him in a very serious tone that his "behavior [was] not accepted at Noisebridge." The young tinkerer looked distressed, keeping his head down to avoid eye contact as another member reprimanded him for "asking people to shut up." A small committee was formed promptly to mediate the conflict, first to talk in private with the fledgling hacker (after the "private talk" design pattern) and then with a bigger group that included people who felt offended by his attitude. After this meeting, the young inventor was never again seen at the space. While his technical experiment was highly welcomed, the disruption his aggressive response caused was in clear violation of the general orientation toward excellence. This rather banal instance short-circuited the moral circuitry of Noisebridge: it was not possible to continue to operate properly as a collective without isolating and addressing the issue.

Another telling instance of do-ocractic practice involved tearing out a section of wall to install a glass window. In an ordinary late afternoon, one of the members with the help of a couple of nonmembers decided to do something about the recurrent theft of tools from the dirty shop. According to them, the problem involved the unhoused contingent of the city allegedly going straight from the main entrance to the dirty shop to peruse the tool cabinets. In order to make it harder for them to leave the space unnoticed, a glass panel was to replace section of wall so that people readily could see who was in the shop. Tearing down the wall did not require collective permission from Noisebridge members. The infrastructural change was immediately implemented and then communicated at the next meeting. The project was ultimately praised, but it faced initial skepticism from members who were wondering why there was a hole in the wall of the dirty shop and an unusual mess at the entrance of the hackerspace. The same member who volunteered to open up the wall could be found regularly helping with other infrastructural matters around the space; he had no "business with computers," he would say, but was always eager to help organize and clean the space. I joined him on one occasion to clean the glass doors by the entrance and remove all the residue of paint and dirt accumulated from long exposure to the grit from the heavy traffic on Mission Street. He was a mystery to many members since he had no interest in "hacking," even in the most expansive sense of the term. He kept coming back and helping the space, I realized, because he found community.

The most active discussion list, "Noisebridge-discuss," was used for general remote communication among members and nonmembers. At times it converged with and indexed the activities of the space, but most of the time it had its own dynamics that hardly reflected the orientation of excellence. In comparison to other hackerspaces in which the mailing list was part of the everyday life of the collective, for Noisebridge there was a stark discontinuity at times. The list, like the Internet Relay Chat (IRC) channel, #noisebridge on Freenode, was mostly populated by people who were interested in the space but who were no longer regular visitors or members. Some of them were, in fact, quite disgruntled with the state of the space. The list was sometimes referred to as "Noisebridge-disgust" because of the intensity of arguments that rapidly escalated into flame wars. The list also served as a space for people to express political positions and address major issues. Feminist members, for example, used the medium to come out publicly about recurrent events of harassment, including not only verbal but also physical harassment against gender minorities. The online space also served as a forum for more practical discussions about what to do with furniture, how to coordinate equipment donation, or to report on the status of the space, announce new projects, fundraising activities, and more. Since the list was generally the place where announcements of disruptive events occurred, it soon became the primary target of internet memes about the status of Noisebridge as a place for "homeless and Arduinos" or the "hippie hackerspace" among other spaces of the global hackerspace network.

One of the technologists I came to interact with at Noisebridge was a GNU/Linux developer who I came to learn a few months later was being accused of being a sexual harasser. His criminal record for domestic violence was published on the Geek Feminism Wiki, a popular community resource. Before his case was made public and notorious in the mailing list, the developer had been well regarded by the community. One day he came to the space and yelled my name across the hackitorium. I flashed him a high five from my seat in front of my computer. I was reading the Noisebridge wiki to remind myself how to operate the laser cutter. (As pieces of infrastructure become part of hackerspaces, people often create public documentation to assist in the process of self-directed learning.) I had been to the training event on how to use the equipment, but I had forgotten the details since I had not put what I had learned to use. After saying hello, the technologist called attention to the stickers on my Thinkpad: Tokyo Hacker Space (THS) and Free Software Foundation Latin America (FSF-LA). He then commented on the Debian conference T-shirt I was wearing. I suddenly realized that I was in "full uniform" or, anthropologically speaking, surrounded by diacritic elements that would call the attention of coders. He did not hesitate to ask me about the shirt, and I mentioned that I volunteered for the conference since I had experience with Free Software–based video streaming. He started to work on a public computer attached to the laser cutter—writing code and building development tools to extend Linux support to an ARM-based computer platform. It was heavy lifting. We had a quick exchange about single-board and low-power computers, and he quipped about the bad aspects of popular projects in this area for their lack of "freedom." As our conversation died down, he started to yell at the computer at every failed code compilation attempt. When I approached his screen to see what was going on, he responded by paraphrasing Thomas Edison in a self-mocking tone: "I am not failing. I am just discovering a thousand ways things should not be done." On a daily basis, he could be found in high spirits around the space, always smiling and interacting with people, having lively conversations about programming and free operating systems. I came to learn after he stopped coming to the space that he was homeless and was being hosted by another nonmember in San Francisco—the same person who would later circulate the information of his past as a sexual harasser. After the discussion of his case in a meeting, he was officially banned: forbidden to enter the space and to send messages to the mailing list. His picture with a description of why he was banned was added to the infamous Noisebridge wiki page 86, listing publicly the personae non gratae for the community.

It was common at Noisebridge for the notion of excellence to be rearticulated through different understandings of hacking but also to be recontextualized after disruptive events. The space held interminable discussions on what it means to hack and to be a hacker, which were usually about how far the concept of excellence could be extended. The attempt at extending the notion of hacking soon became the source of jokes by members, for instance, in the discussions of "food hacking" as people disputed the merit of using the term for alternative methods of food preparation (or even for sleeping at Noisebridge under the ironic guise of "sleep hacking"). In one of the recurrent conflicts between the "sleep hackers" and hackers with systems administration background, domain-name resolution for Facebook.com was blocked for the local network as a trolling campaign against the perceived abuse of the space. It did not take long before a complaint was sent to the mailing list: "Facebook turns off after 11pm. I'll admit it was a little funny but it interferes with my work. Please remove it." The reply came shortly: "If you can't figure out how to overcome DNS-based blocking (hint: no outbound packet filtering/firewalling is taking place), then you probably shouldn't be hanging out at Noisebridge in the middle of the night."[15]

Once I was entering the space and noticed a whiteboard next to the gate with the question "What is hacking?" Around the question I could read disputing definitions written in different peoples' handwriting. One of the newcomers, an Occupy activist who had learned about the space through the Homes Not Jails collective, articulated his understanding in the following terms: "Hackers are people who find clever & subversive ways to do things outside the system, often in defiance of the establishment & in service of revolutionary or anarchistic principles." He then took it back, adding an important parenthesis: "not always. Some hackers are apolitical, insofar as it is possible to be so these days; I have also met hackers who seemed to me to be quite conservative politically."[16]

Mitch was keen on this expansive notion of hacking as well. In a presentation at an unconference in San Francisco, he proposed that hacking was "taking anything in the world, using it as a resource, doing cool things with it, learning from it, seeing what happens, what works, what does not, and sharing it. But, above all, doing it because we love it. We can hack anything. Of course, people think of computers and tech, but we can also hack ourselves, our lives, our resources." This definition encompassed a wide range of commoning practices around Noisebridge that were not computer- or electronics-based—and certainly one of the most powerful aspects of the collective experience. It also pointed to the demand that was embedded in the notion of hacking as a moral orientation for self-directed learning: by engaging in experimental practices with various resources in the world (a resourceful and highly privileged world for some, it is important to say), one was encouraged to situate oneself in a position to improve one's life and that of those around them. The parallel here with the practices of neoliberal self-making are not incidental but cannot be conflated because there is one fundamental and irreconcilable difference: hackerspaces are meant, first and foremost, for convivial technologies that are diametrically opposed to the utilitarian logic of the entrepreneurial self.

Hacking, broadly defined, was fostered at Noisebridge as one of the main guiding forces for supporting conviviality in organizational, political, and material terms. Open-ended definitions of hacking, albeit disputed, served for encounters across difference that were not accommodated by other tech spaces. Excellence worked as an operator of commoning—that is, it was meant to modulate relations of exchange between people and among people and technical objects as open projects that involved, for instance, "hacking for the collective good" (with book scanners to make books available digitally) or "hacking for political causes" (by hosting fast nodes of the Tor network for fostering privacy protection). Right before entering the space, one could not help reading the warning by the doorsteps: "You are responsible for the vibes you bring here, be excellent." Conflicts surfaced whenever different groups with different perceptions of the purposes of the hackerspace were confronted: differences were marked in their bodily *hexis*—differences of socioeconomic background, formal education and technical cultivation, gender identification, and ethnic, national, and racial belonging. These differences accounted for tensions in horizontalist governance.

Performing Horizontal Governance

During Noisebridge's first year, before moving into their first space, a large but close-knit group of founders held face-to-face weekly meetings to explore options for the organization of collective decision-making and conflict resolution. They decided to apply for legal status as a tax-exempt educational nonprofit membership organization under the Internal Revenue Code 501(c)(3) of the US tax statutory law. Technologists with experience in other hackerspaces and professional associations chimed in, offering suggestions about the role of the board of directors, their prerogatives, and those of the members. There was a tacit agreement that the board would not be more than illustrative and therefore not having any oversight of members' projects or governance of the space. The board would be appointed only to fulfill a legal requirement, not an organizational need. The running joke was that the board members could be easily substituted by a few lines of a Unix shell script. During the early meetings, there was no perceived need to create a set of rules, but there was a strong sense of the importance of having a consensus-based decision-making process. It was decided the collective should have only one rule, subjective by design: "Be excellent to each other," whose meaning would continually evolve and morph with ongoing discussions on what it means and does not mean. Legally, the board had responsibility over Noisebridge, but in practice its members were not differentiated from other members. As a mechanism to protect the board members, insurance was paid for each position (president, secretary, and treasurer) with resources that were obtained through membership dues and lots of small donations from individuals, plus regular donations from Bay Area IT professionals (along with matching donations from their employers, again, mainly driven by tax exemptions). Formal positions of authority were commonly made fun of by members. One former president of Noisebridge put it like this: "It was my understanding that the meeting got to a point where everyone agreed that the next person to walk through the door should be the president. . . . I walked through the door next." An influential member, well-known in the community for his anarchist literary inclinations and digital rights activism, composed a verse that thoughtfully captures this cardinal orientation of horizontal governance:

> For we're excellent to each other here
> We rarely ever block
> We value tools over pre-emptive rules
> And spurn the key and the lock.
> —malaclyps[17]

Following the "regularity design patterns," Noisebridge held Tuesday meetings that were not only aspirationally but effectively open to anyone. Meetings were important spaces for debates where tensions surfaced, but also for the introduction of members, nonmembers, and aspiring members with particular interests and skills. They were also a forum for self-introductions, sometimes with performances of expertise, where both members and newcomers had the chance to connect with others, make an impression, and find collaborators. Participants would preface their self-introduction with short affirmative sentences, such as "I hack electronics" or "I hack Go" (programming language). Regular members would also take the opportunity to joke about their own area of expertise or interest. Many would refrain from talking about themselves; they would instead just say their first name or IRC nickname followed by an area of focus (such as "working on censorship detection," "making a radiation detector," "interested in software-defined radio," "building a startup," "game programming," "signal processing," "lockpicking," or "just curious").

Weekly meeting rites included a brief description of what Noisebridge is about, which was a way of reinforcing it for members while describing it to the newcomers. "Does anyone want to say what Noisebridge is about?" was the most common opening question. The moderator and notetaker for each meeting were chosen on the spot, usually taken up by members or nonmembers who were familiar with the meeting protocol. Another recurrent question often asked for pedagogical purposes was "Can someone describe the Noisebridge consensus process?" This was followed by a ritual reciting of what the space was and what consensus meant—also known by the name Consenso, an AI imagined and popularized by Liz Henry and Danny O'Brien. At times, a newcomer would take the liberty of articulating their own version. "It means," one aspiring member once said, "the ability of an individual to go against the collective." One of the seasoned members immediately rectified that definition: "I think someone is confused about what consensus is and what it is for. Does anyone want to explain?" These were ordinary but key pedagogical instances whose efficacy relied upon the fact that they were not framed as such. They gave visibility to the moral circuitry of the collective.

Noisebridge's governance was sustained as horizontally as possible, despite occasional conflicts stemming from differences of background, which were recognized and displayed through different forms of technical competence with consensus-based decision-making. Participants preferred to avoid differentiating between members and nonmembers, but the distinction was often foregrounded in moments of crisis over consensus items—which were religiously archived on the wiki and discussed in and outside meetings until a consensual position was agreed upon. Decisions could be blocked in extreme cases via proxy, a person permitted to speak on behalf of an absent member. Members could also "stand aside" when no consensus could be reached. Formally, members were distinguished from nonmembers in two ways: members had monthly membership dues (which could be skipped in cases of financial hardship), and they could block consensus. Membership, however, was not trivial to obtain. As it was usually explained at meetings, it was not "the same as membership in a gym or a chess club." Since people without "a proper understanding of consensus" could hamper the space's decision-making process, members were very careful to protect the collective from accepting a person who could make the process more difficult. Membership took into account contributions to the community and, quite often, assessments of technical abilities. Considerations of previous experiences with horizontal politics and personal connections were also key in accepting a new member.

Positions of power were mocked as legal formalities, but the process to fulfill legal obligations was carried out. Board members were voted in after members were picked as candidates. The whole voting process was taken very lightly, and the positions were not perceived as entitlement above other members' and nonmembers' voices, but they tended to be occupied by members with a certain degree of prestige. In one lively and memorable occasion of voting, ballots were made into paper balls and tossed around as they would be in an all-out classroom paper-ball war. Members consulted with others on the spot regarding the voting rules; informally, candidates were picked and votes were cast in a joking manner. Every obligatory vote was mocked and served as an opportunity for animating this anti-structural moment for the collective—a veritable anti-representative politics machine in action.

Membership in the hackerspace, on the other hand, was taken very seriously. It could be obtained through a process of application and consensual deliberation among participants at a meeting in which an applicant was presented as a candidate. Members could veto an application in person or via proxy. Applicants had to collect the signature of three members in good standing and write about what brought them to Noisebridge, an important narrative that usually glossed over saturated symbols such as "hacking," "community," and "collaboration," identifying the hackerspace as a commons (not a "club good"). Applications were archived in a binder that was available for everyone to see and kept by the entrance of the space. Any new applications were read at each meeting to be considered by the present attendees. If the applicant was not present, the application would be held for the next meeting and the following three weeks before it was discarded and had to be reinitiated. If an application contained three signatures and if the applicant was present, the application was read and the applicant was invited to leave Noisebridge so that the present members and nonmembers could discuss any issues they might have had with the person. Upon return, the applicant was supposed to bring gifts, something small but appreciated by the group, such as a six-pack of beer. This obligation was usually communicated in a playful tone.

The membership process was a ritual in its own right with regular phases to demarcate the passage of status: *separation*, *liminality*, and, finally, *integration*.[18] The temporary separation of the applicant generated feelings of anxiety, apprehension, and nervousness. This period would generally come to an end whenever a consensual position regarding the application was reached. Any opposition to the application had to be voiced and justified. Arguments in favor of the applicant tended to be about their technical abilities, capacity to get along with others, consensus-making abilities, and projects they were involved with at the space. After a series of questions to probe if the applicant was trustworthy and committed— "In what circumstances would you block consensus?" or "What is your understanding of the consensus process?"—the applicant had a chance to respond, followed by a period of evaluation of the responses. Arguments against the acceptance of an application were voiced during the absence of the candidate, but they were also formulated as a series of probing questions upon the candidate's return to the meeting.

The whole session served not only as a ritual for incorporation of a new member but also as a pedagogical moment for nonmembers, with notions of what consensus and excellence meant being voiced and, if needed, corrected with the version that members encoded in the wiki. Applicants were also asked about controversial topics, such as removing the kitchen in order to stop attracting "people who were not at Noisebridge to hack" or having locking hours to prevent people from sleeping at the space. During the Q and A, the applicant would usually try to respond by mobilizing preconstructed elements from discourses the applicant had been exposed to around Noisebridge. For example, applicants who were not very familiar with consensus-based decision-making usually started to learn by attending meetings before the evaluation of their application and listening to more experienced members' assessments, such as "that person is a little bit confused about how things work around here."

The aspiring member who had said that *consensus* meant "the ability of the individual to go against collective decisions" faced strong resistance during the evaluation of his candidacy. After the entire ritual had come to an end, I found the newly accepted member outside the space, looking relieved and having a conversation with nonmembers. We were approaching midnight, and soon I would not be able to catch a bus to return home in the Upper Haight. Speaking of his acceptance, he confided, "If I could hack Google like that to become a member, they would hire me, just because I hacked my way in." After a shaky exchange regarding the nature and purpose of consensus, he had managed to save his membership application somehow by talking to the group about his ideas on computing and neurobiology. For a group of nonexperts, his discussion of the computer/brain interface seemed curiously well-informed. No one (which is unusual when someone talks about technoscientific topics at Noisebridge) challenged his goals and observations. It was followed by a period of silence. On my way out of the space, I congratulated him by saying it must have been great to be accepted after all. He nodded with a huge smile, surrounded by a cloud of smoke, with glowing red cheeks. He seemed relieved after being "grilled" by the collective in a crowded meeting.

Another mechanism to control participation in the space that emerged over time was asking people to leave, which had not only become much more frequent after Occupy San Francisco, but was eventually formalized as a procedure. People who are considered disruptive or perceived as an abuser of others and of the space (for overuse such as sleeping, only cooking, or spending "too much time" idling around) can be asked to leave. Those who are subjected to this informal policy, as a last resort, are supposed to return for a Tuesday meeting to discuss their behavior and, if the matter is serious enough, they can be banned permanently, which then becomes a "consensus item," after which it goes into immediate effect and is published on the wiki and the mailing lists. Critical events involving sexual harassment, theft, or violence have often resulted in the permanent banning of nonmembers. Interestingly, no member has ever been banned.

Occupy San Francisco and Hackupy

By late 2011, one year after the onset of the Occupy movement in San Francisco, a heated discussion revolved around the "Occupy group" at an ordinary Tuesday meeting. One of the critical voices suggested that the group attracted a lot of people they "did not want at Noisebridge." There were complaints about "the space not having hackers anymore," suggesting its "decaying state when it used to be great." The Occupy movement itself was not the topic of contention. As one of the founder put it, "Many people who lived in the streets were welcome because they *do belong* at Noisebridge." But as the "open-door" experiment unfolded, the radical openness of the collective was put to test. Anthropologists who engaged in Occupy have described the ways in which the movement organized spaces of contestation with prefigurative sociality in self-organized initiatives.[19] In the context of the early mobilizations of September 2011, Noisebridge was an obvious ally in providing technologies to facilitate virtual and actual gatherings, bridging digital technologies and political action until more pronounced differences in socioeconomic conditions surfaced. Before Occupy SF, there were numerous intersectional bridges of members' personal trajectories with autonomist groups. Activities were dedicated to gender minorities in computing, and affinity groups gathered around the issues of e-waste recycling, time-banking, and "hacking mobility" (hack-ability) for members with disabilities. Alongside Occupy camps, Noisebridge was a place of aggregation and conviviality. But over time, solidarity was converted into strong rejection.

I spent my first night working at Noisebridge after a couple months of regular attendance, working alongside one of the nonmembers who was closely identified with the Occupy movement. The same group could be found throughout the day and evening at the space, some sitting quietly at one of the public computers, others sleeping on the couch by the library. I would come to learn the stories of some of them on the fateful day they were to be banned due to accusations of "non-excellent behavior." I ended up working with a person who was eager to learn the GNU/Linux operating system and live video streaming, two technical topics I had some experience to share. We found an old rack server at the "hacker alley" (a wall full of shelves with free electronics and computer parts) and decided to build a streaming server for testing. My partner in the project was eager to peruse piles and piles of electronic junk for missing parts, RAM modules, and usable hard disks. We spent much of the night talking about use cases for our nascent project: we could provide a live video feed and let other people participate in meetings and activities taking place at Noisebridge, or we could provide activists with the same service during demonstrations. At the time, I pitched the idea of a portable audio streamer based on the popular Raspberry Pi single-board computer—baptizing it "Raspbarricade" under the influence of the political moment. I was drawn to the technical project mostly for the possibility of exchange, but also due to its potential political usage and the novelty of an open implementation of new video codecs and streaming protocols that I wanted to try out. I explained the details of what I was doing to my co-participant while building the experimental, alpha version software we needed for the server. We set up our working area in a more secluded corner surrounded by disassembled personal computers and hidden from the curious gaze of Noisebridge members, against a wall of old electronics testing equipment. We had everything we needed at arm's length: an analog video camera, a video capture card, a fairly robust server that we had to reassemble from scratch, and plenty of network bandwidth for testing. My partner in the experiment— himself a regular at Noisebridge and a squatter in San Francisco—casually shared stories of his youth and those of other nonmembers in the space: their struggle with drugs, their deep engagement with the Occupy movement, and the fact that it was harder for them to get jobs because they did not have a place to sleep or shower. He talked about his early experiences with a Commodore 64 and the tricks he and his friends had learned to perform in order to break copy protection of games. He had a strong but remote connection with hacking.

In one of the many contentious meetings where people would complain about the state of the space—essentially that the facilities were dirty and nobody seemed to care—a member went off blaming some of the nonmembers for their homelessness: "They spent a year at Noisebridge," scolding no one in particular while leveling his critique at everyone. "They could have learned something to get a job, but they didn't." Upon my return the next day to continue working on the server, I found the hard disk was missing. My partner told me he had procured a RAM module that had also disappeared, so we both gave up the project right there. There was a shared feeling of disappointment. It was obvious to me that someone took the opportunity to make a few dollars off the parts we salvaged. Noisebridge's tools and equipment were perceived by some as though they were common goods of a public library. To expropriate what was not private property in the first place felt disturbingly wrong, except, of course, in face of the need that was blatantly obvious all around the neighborhood and the city. The need for housing, food, and a livable source of income. We never talked about our experiment again. In the span of two months, my partner in his "present-absence" in the life of the space eventually left without a trace.

When the Occupy movement took San Francisco's downtown with street marches, sit-ins, and public-space occupations, Noisebridge came in to help in various capacities. The space was used to hold meetings for Occupy, discussing activities meant to bridge activists with and without technical backgrounds—the so-called Hackupy meetings. Some of the founders of Noisebridge set up a tent at the Occupy San Francisco camp, providing technical support with pedal-power sources and audio-streaming services and holding hackathons for building web pages for the movement, among other actions. My partner in the Free Software–based video-streaming experiment was a permanent fixture in the Occupy SF camp. Hackupy was one of the short-lived initiatives to connect hackerspaces with the movement, similar to the role fulfilled by "hacklabs" a decade earlier in the context of the alterglobalization movement with the international expansion of Independent Media Centers.[20] An active Noisebridge member said this of the experience: "My hopes are that more hacker spaces will jump in on this and list out their hacker space as open to those in need of help. . . . At Noisebridge we've had people hacking away on Occupy tech more or less every night."

The intersection of Noisebridge with the Occupy movement was eventually perceived as a major source of disruption. As the space grew in popularity, the influx of strangers also grew considerably. Before the onset of Occupy, there were numerous intersectional bridges across members' personal trajectories with social movements and autonomist forms of political organization. Workshops were dedicated to gender minorities and there were activities around hacking mobility for members with disabilities and special needs. The solidarity and active participation in the Occupy movement was felt as a natural progression for most members, given their political commitment to the common, one of the main reasons why they were drawn to participate in creating and sustaining Noisebridge in the first place. What did not work as the experience of exchange with Occupy activists unfolded was the felt difference in the context of the everyday life of the movement's actions between more established middle-class lifestyles, represented mostly by some of the members with high-paying tech jobs, in contrast to activist groups, which combined groups of housing justice and unhoused activists. The ulterior occupation of Noisebridge altered the perception of the space with respect to its openness, frustrating founding members who felt abused in their position as allies. One of the founders described this issue to me in the following terms: "There were too many people at Noisebridge who did not care about the Noisebridge community, who didn't care about others who were at Noisebridge before them, who stole things, destroyed things, shot up drugs, hit on women, slept in the space, took up space, and did not contribute in any way. This drove away vast numbers of people who did care and who used to contribute. Noisebridge did not exist for such activities."

In the process of disoccupation of downtown San Francisco, Noisebridge started to receive a very particular contingent of activists, identified by some members as "freeloaders or oogles"—in street slang, youth of middle-class backgrounds who were said to have become homeless by choice rather than necessity. One of the founding members lamented the state of the space at the time: "It's becoming increasingly clear to me that Noisebridge has jumped the shark somewhat. The people I find myself most running into there are not into technology, hacking, or, really, Noisebridge. I have found them to be mostly belligerent, uninterested in rational conversations, and seemingly trying to push various political agendas. I am getting tired of seeing the hackerspace that I worked so hard to help build fall apart, both socially and physically." Another founding member said he had joined because he liked the idea of a club for specialists to study computing. "Groups in Europe," he observed, "are closed for nonmembers and hard to enter." He then qualified that statement by adding that he did not value exclusivity, he only thought that the radical openness of Noisebridge made it lose its focus and purpose. "It is not a place for people who are experts to work on projects, but a general space for people with loose affiliation with the idea of hacking, computing, activism, etc." He also suggested that, because of openness and lack of orientation, the community could not defend itself, which was clear in the recurrent cases of theft and assault against women-identified members, as well as in the difficulty the collective had in preventing people from sleeping at Noisebridge and returning to the space after being banned. The article "The Tyranny of Structurelessness" by Jo Freeman[21] had a big influence on him in its articulation of how "tyrannical structures" take shape when there is an assumption of structurelessness; in other words, in the absence of explicit norms, crypto-authoritarian structures of power tend to rule social relations and mechanisms of collective self-governance.

A couple of years after Occupy SF, unhoused activists kept returning to Noisebridge only to be invited to leave the space again or to be banned for the perceived overuse and abuse of the space, its people, and its resources. They were not banned because they were unhoused, but, in the words of a member who experienced this issue firsthand, "because they were not contributing, not caring, doing things that were incredibly unexcellent to others." I was told by another member during an interview that it had all become too much and it was driving people away, burning them out, and frustrating their expectations for the space:

> There was some point with Occupy—this is hard to say because I feel I am betraying my ideals—but I got so frustrated that people were living [at NB]. I got so frustrated that people get out of jail and come hang out at Noisebridge. I was "Oh, great!" I agree with rehabilitating people. It is not that I would not hire people because they had been in jail, but there is a way to say "Hey, you are so much drama, you are so dysfunctional that you can't be here because you are driving away people who can function." We reached a point where we had to have an emergency phone for the night, because the people there at night were too afraid to call the police, so they had to call someone middle-class like us to be the front man, because they were all on parole, have immigration status. Oh, this is not good! You cannot really have a middle-class person you call to solve these problems, but we did for about a year during Occupy.

The solidarity ties that were cultivated during Occupy could not be sustained over time. At the crux of the matter, there was a lack of exchange ties between unhoused activists and technologists, exchanges of technopolitical nature that were fundamental for sustaining the particular moral circuitry of the hackerspace.

Supporting the Hackerspace from a Distance

With few exceptions, the original group of founders could no longer be found at Noisebridge, but many of them still kept their membership status and helped to maintain the network infrastructure. One of these founding members, Andy Isaacson, who assisted the space remotely, summarized his trajectory as "growing up in mid-country, raised by Republican parents, who took seriously the compassion dimension of Protestantism." He had memories of doing interesting things with computers early on and of growing up "modestly but never had any economic hardship." His parents were very strict and he never developed any interest in the computer experiments that were classified as "digital trespassing" after the Computer Fraud and Abuse Act (CFAA) was passed. Following the laws because "they were rational," he reminisced, had been his take on the question of criminalization of hacking. "Not all laws are good," he would add, "but there are ways to break laws while keeping on par with the spirit of laws."

His career in information security started when he discovered cryptography by learning about Pretty Good Privacy (PGP) and the "Crypto Wars" from *Wired* magazine in the mid-1990s—an important event in the history of hacking involving a series of manifestos and protests against the governmental restriction of civilian use and distribution of cryptographic technologies. At that time, Andy also was exposed to a serious dose of "Ayn Rand-libertarianism coming from white males from the Bay and the Valley," which characterized well the ultraliberal leaning of the most vocal cypherpunks. His English literature teacher introduced him to Ayn Rand, and he dove into her works only to rethink his position later in life. His early transformative experiences involved connecting a small ISP to the internet and exploiting security flaws in his homework assignment in college. With a friend, he reverse engineered a piece of software that had been distributed by his professor in order to show how it allowed for privilege escalation. His instructor was not pleased with his finding. He could not find professors and researchers at Michigan Tech to help him with cryptography and security research at the time, so he proceeded with his experiments independently. He wrote a script to log into the machines at the computer lab when they were idle and use their computing power to help in a challenge to break a cypher; his college contributed 10 percent of the total computational power used to eventually meet the challenge. That was an exhilarating experience. After college, he started working for BitMover (on a source code version control system called BitKeeper). He was later recommended for a job at Broadcom, which hired him with a relocation package to the Bay Area to work on embedded systems development, porting the Linux kernel to an MIPS32-based chip.

Among his political experiences, Andy cites participating in the debate team of his high school, but in his words, their outlook was "connected to a very traditional, American mainstream of commercial newspapers as sources for debate among Republican and Democrat candidates." He did not identify with a particular political tradition. After starting out much closer to North American libertarianism, he changed his views substantially as he got older. He believes in "individual freedom and the importance of working toward it," which is something that "fundamentally moves him to work in information security." He was one of the maintainers of the influential Noisetor project, contributing to the management of one of the fastest "exit nodes" in the network. He was introduced to Tor by another Noisebridge founder and internet privacy activist, who would later cause a huge commotion in the community due to accusations against him of sexual abuse. Early on, he and Andy decided to work together on a fundraising project to pay for the bandwidth costs of powerful nodes and collaborate with a European consortium to support the Tor network on a global scale. It worked amazingly well, and Noisebridge became a powerful host for a good number of years.

Andy was very active in the early years of Noisebridge but progressively moved away from the space. The promise he saw in Noisebridge was fueled by the excitement he experienced at the Chaos Computer Camp in 2007, which gave rise to a wave of new hackerspaces. He was so committed to Noisebridge that, at some point, he got burned out. Excellence, do-ocracy, and consensus were basic values that had been present since its inception. They were adopted, in part, because some of the founders were adamant in supporting ideas of radical openness and horizontality. "Due to their charisma, the acceptance of these ideas was not problematized," according to him, "but accepted in good faith." They were the basis of an experiment, after all, but in his evaluation, it did not quite work. He confided he was very disappointed with the state of the space and rarely came to visit, helping only from afar.

Actual and Imagined Autonomy

Noisebridge is more of a prefigurative experiment than an accomplishment of horizontalist values, even though it is possible to identify efforts to guide interactions with values of excellence, autonomy, and do-ocracy throughout its history. These values have been the shared horizon of an alternative technopolitical present, for they constituted an actual space where dispositions for commoning could be cultivated. This cultivation, however, was not without intersectional frictions relating to class, gender, and technical expertise. Horizontality was one of the key sources of debate and concern. Meetings extended over long hours, and some decisions seemed, at times, almost impossible to implement, such as the proposal to control access to the space. The collective short-circuited more often than not when different elements were introduced.

Noisebridge was imagined as a symbol of autonomist tech development, but it was also the object of skepticism and scorn from other hackerspaces due to its radical aspiration: to offer an alternative space for commoning where IT labor regimes were to be questioned. Noisebridge's core symbols of excellence, openness, and horizontal governance were articulated differently and disputed as divergent groups of specialists, entrepreneurial figures, political activists, and curious technologists came to share the space.

In one of the innumerable exchanges I had about single-board computers, which I was experimenting with at the time, a startup founder and active member of Noisebridge told me he had acquired his boards "not for play, but for product development," so he "couldn't show [them to me] until the release of a new startup product." Another telling instance of interdiction happened at Circuit Hacking Monday. We were all busy in an ordinary hardware workshop when a young woman in her mid-twenties approached the organizer to ask for one of his kits. She did not yet know how to solder and affirmed she was there to learn. The instructor confirmed she was at the right place, with a broad, welcoming smile. They exchanged the kit for twenty US dollars. In about five minutes of dedicated hands-on demonstration—which the instructor patiently repeated for a second time after his demonstration for the whole group—she was ready to start assembling. She opened up her laptop, which had curious blue keys contrasting with other keys. When asked about those colored keys, she said it was her invention and she could not talk about it. The organizer kept to himself, dismayed. An awkward silence ensued. People kept their heads down, concentrating on their solder joints, including me as I sat next to her, practicing surface-mount soldering.

Despite the general orientation toward openness and sharing, monetary and voluntary transactions on the register of the market and the gift intertwined. The purchase of a kit was not mandatory, as any nonmember was welcomed to bring their own projects to the table, as I had done that day. Oftentimes, engineers with small electronics businesses would come to help people with questions. At times, they would offer their own products and kits. The space and time of the workshop, which extended itself from three to four hours on average, occupying a good part of the evening, was organized in such a way to guarantee that interactions happened mostly under an open register, but other hegemonic values often intervened, "No, I can't talk about or show you my project" was a fairly common interdiction in the context of the tech industry under the ubiquitous injunction of non-disclosure agreements (NDAs). For the moral circuitry of the hackerspace, however, ties were created among co-participants by generating a debt on the register of the gift: the demonstration of how to accomplish a certain technical task was a gift of dedication that opened up a whole series of technical objects. As Open Hardware projects were distributed with their schematics, design files, and bill of materials, it was possible to re-create them without extra effort or cost other than the market price for the materials. Having been given the possibility of replication, social and economic ties could be made, reactivated, or changed. But this cycle could just as easily be broken, disrupting the circuit of contributions in the economy of public workshops with open technologies. When participants refused to share information about their project or when a workshop participant refused to speak about their creation while they learned from someone else's, sociotechnical ties and future cycles of exchange (which followed an initial donation) were foreclosed.

Grounded in the context of practical exchange within a specific moral circuitry, the imaginative force surrounding Noisebridge was exerted until recently through narratives of Open Hardware hacking as a model for new projects in new community spaces. Its place as an imagined space where all the best projects and the most brilliant hackers gathered has been attacked as outsiders have come to characterize Noisebridge as a "toxic place." The hackerspace, which offered a window into a wide spectrum of socioeconomic lives and technopolitical orientations in the San Francisco Bay Area, was unique in its possibilities and difficulties, as we will see more clearly when we examine the collective experiences of other hackerspaces in the following chapters.

The prefigurative political experiments at Noisebridge were eventually challenged when the orientation toward radical openness, horizontality, and excellence was met with some of the most precarious constituents of

Occupy—among those who would not engage in technical exchange but would use the space without contributing to it. In the wake of its so-called dark age, Noisebridge members felt a disruption in their organizational forms and ultimately in their convivial mode of exchange. Their solution was to propose a "reboot" for six months, when the only people allowed in were those who were cleaning, fixing, painting, reorganizing, and discussing how to move forward with the hackerspace. The principle of radical openness was temporarily suspended through the process, temporarily establishing forms of regulation for "keyed members," a new control measure used by most spaces in the hackerspace network: spaces were made accessible to the public for Tuesday meetings (according to the "hackerspace design patterns") but access to the space otherwise followed restrictive norms. After the reboot, Noisebridge returned to its original state as a space of transnational exchange and place of technopolitical experience with one change to the radical openness that had characterized its first six years: the door would be locked, but anyone without a key or door code could ring the doorbell. Someone who let a person in took responsibility for the new person and was obliged to give them a tour if they were not recognized. The tour included information on what the community was about, the one and only rule of excellence, the consensus-driven decision-making process, and do-ocracy—all in the tour guide's own words. Another key outcome of the reboot was the subjective articulation of who belonged at Noisebridge: someone who both *benefited from* and *contributed to* the community while following the one rule.

In 2020, Noisebridge moved to another warehouse on the same block, and as of this writing, it is still open and going strong with the similar sociotechnical openness. The community has proven to be incredibly resilient. The IBM XT that I brought to the space a decade ago is still decorating its entrance and running the same BASIC routine someone do-ocratically programmed. But now the work *HACK* is permanently etched into its monitor's phosphors after running the same code uninterruptedly for many years. When asked whether Noisebridge would qualify as an "anarchist hackerspace," long-term member and agitator Malaclyps got to crux of the experimental matter: "Not until everywhere else is an anarchist hackerspace too! Or, at least, that is the plan . . ."[22]

TWO

Altman

Yearning for Community

At the main stage of the 14th International Free Software Forum (FISL) in Porto Alegre, Mitch Altman delivered a talk after our tour of Brazilian hackerspaces, where he taught newcomers to assemble kits just like the ones he used to introduce tens of thousands of Noisebridge visitors over the years. In an upbeat tone, his talk was playful, exalted, and charged with charismatic flair. During the Q and A session, Free Software hacker and activist Richard Stallman appeared out of the blue in the back of the conference room and questioned him about the usage of the term *Open Source*. Stallman acknowledged Mitch's political commitment and lamented the fact that he was using the term to defend software and hardware "freedom." In his talk, Mitch had primarily questioned meaningless jobs in the IT industry by pointing to the fact that he was making a living with open technologies that were meaningful for him and that it was possible for other technologists to do the same. He laid out his vision of technological production based on convivial, anti-militaristic ideals and shared an opinion he repeatedly invoked in his talks, interviews, and workshops around the world: *I am really glad people are fighting against horrible things perpetrated by institutions, corporations, and individuals. My approach is to en-*

vision the positive that I want and attract more people, putting the word out to people who jive with that positive vision—that makes it stronger, that attracts more people. As I see it, it is not primarily about resources, even though they are important. It is not about money, although it is also important. It is about creating opportunities for more people to live fulfilling lives.

Mitch was born in a family of Chicago activists. His grandfather was a labor activist and communist sympathizer in the 1930s who had lost his tailor shop in an arson attack following the Mafia takeover of the Teamsters Union. His parents got involved with the counterculture movement when Mitch was young. His mother was an elementary school teacher. His father, who had terrible experiences in a clerical position in the US Army as a communications officer before Mitch's birth, had started a successful architecture firm. *My father loved architecture and art, so he arranged his life such that he would have a lot of time for it. I think that is a big part of why I do not have what seems to be the disease of the modern world of focusing on money. It is not the money that makes us happy. To have the life you want to live is beyond subsistence. It is not about the resources; it's about living the life you want to live. We did not have a lot growing up. Still, we were certainly privileged for being white in America, white and middle class, albeit lower-middle class.*

Mitch had been encouraged to explore his material surroundings when he was a child. As the standard narrative goes for most engineers, he also started taking things apart when he was very young, learning from popular science and electronics magazines, and creating his own toys. He once dreamt of pushing his mother's vacuum-tube radio off the counter to see its internal parts. Eventually, the radio mysteriously fell from the table and smashed on the floor to reveal its innards. Mitch was six when this happened, marking his earliest realization of the dream of exploring electronics. From this moment on, electronics became his constant companion, as he landed jobs, contracts, and multiple invitations to lead electronics workshops in various parts of the world.

Mitch's mother was a leader of a Cub Scout troop, which he took part in as a child. One of his memorable experiences in the group was having access to a mainframe computer in the 1960s when computers figured mostly in science fiction accounts. He went to visit a Sara Lee factory with his fellow scouts and had the first contact with a computer machine. *The factory made pastries with robots controlled by this mainframe computer and ran twenty-four hours a day, but during the night it had reduced production, and they did not need all the computing power. We would go there to punch cards. It was my first contact with a real computer. We had to learn IBM 360 or 370 as-sembly language; that was all we could program in. They didn't have BASIC. I do not know what languages they had to control the robots, but I think they used assembly—otherwise, why wouldn't they let us use FORTRAN? We were kind of on our own. The computer operator showed us a little bit of what to do, gave us manuals. . . . Boy Scouting is not just idiotic; it is horrible. They teach kids not only to be militarist but to be closed-minded and conformist, to exclude people who are different from them, especially if they are gay. Cub Scouts is part of that same organization, but it is different. It is more about getting together and play-ing games and doing activities for merit badges, which was fun. They did not have any technology for merit badges when I was a kid, but since we were just a bunch of geeky kids, we got lucky* [to have access to a computer at this age]. *Hardly anybody did.*

Growing up, Mitch was surrounded by debates by his parents and their friends about the wrongdoings of the government concerning the Viet-nam War. His parents' friends would come over to his house to discuss political issues, giving Mitch an early exposure. *Throughout the '60s, my folks organized environmental activism, anti-war, and pro-civil rights—all sorts of cool things. Brought up in that environment, I learned that the army really sucks.* He has recollections of discussions about rights and wrongs and being encouraged to think through political issues by himself. *What is this Vietnam war about? Why are we being lied to about it? What is the point of all of this? People are being forced to suffer greatly. And for what? Freedom? If this is about freedom, why are we making these people suffer so much? Is not the point of freedom to allow everyone to live better lives? It was clearly not about free-dom. The real reasons were unspoken. From my point of view as a young person, the right thing to do was to oppose it. Why should we support something that is making the world a worse place? Growing up in that kind of environment, the focus was on doing the right thing and thinking about things on a more political scale and not just about my own life and bank account.* His coming of age in politics happened while George McGovern campaigned against Richard Nixon in 1970. *McGovern was not only liberal but actually progressive (which is as far in that direction as possible for anyone at that level of institutional politics in the US). My parents organized for him, and I was volunteering for him. The CIA, NSA, and others from the intelligence community were monitoring what we were doing (as we found out later through freedom-of-information requests). All of this got me very disillusioned with the whole electoral process. I stopped putting my energies there.*

Mitch's parents had the perception that the public school in their Chicago district was subpar, so they moved to the suburbs. *School was supposed to be better in the suburbs, but I got bullied and beaten up every day there, so I focused on doing geeky things instead of interacting with people.* By retreating into more individual pastimes involving electronics, such as creating an interphone system to connect his bed with his brother's, he got hooked on TV shows, which he considers his very first of a series of addictions: *I was terrified of people, pathologically shy.* In high school, Mitch continued watching TV and found a community with a group of students smoking weed recreationally. *When I was not high, I was terribly depressed, and when I was high, I focused on how depressed I was and how fucked up I was. I had to quit pot. I then went to other drugs and other addictive behaviors. But at the same time, I was focusing on forming communities and found various ways to do that.* His high school had an electronics program, which basically meant training for aspiring TV and radio technicians. There he found a community and had to venture out of himself and beyond the small circle of friends who had only pot in common. His electronics teacher created an exceptional environment for sharing among students and allowed plenty of room for exploration. *Even though I knew more than he did about electronics, he was not threatened by that, and he actually created an environment for me and other kids to play and explore. He encouraged us to share what we knew with each other. Very proto-hackerspace-ish.*

After being bullied regularly in school, going through a life of torment and forced introspection, and finding himself inadequate most of the time, electronics became a safe harbor. The transition from childhood to adolescence immersed in electronics led smoothly to an engineering degree in college. Mitch attended the University of Illinois, Urbana-Champaign where he soon found himself working in the only laboratory that did not accept any military funding. *The lab coordinator was a Chilean electrical engineering professor who fled in exile due to Pinochet coming to power with the help of the US military. When I got there in the mid-1970s he was about to get fired for lack of funding, but Intel, a company that makes computer chips, donated a million dollars worth of microcontrollers to us, so suddenly our lab had an incredible resource that existed nowhere else on the planet.* Mitch soon became a lab assistant and taught various classes using microcontrollers that were suddenly widely available. *I ended up being a teaching assistant in graduate school. I tried to make my classes—as much as possible—more than just me in front of students imparting knowledge and information, but instead giving people enough knowledge and fun tasks to do and letting them go do it, so that I was just a resource and a mentor at that point. It worked really well. This is what I did later in life by helping people come together and helping them form communities.* Before advancing to graduate school in electronics engineering, he took courses in psychology and anthropology, which fascinated him. He recalls asking himself how people work and pursuing questions regarding the unconscious and the nature of human psychological drives. *While all the other labs were doing guided-missile systems and other stupid things, in our lab, without military funding, we were doing more interesting projects, such as robots, flying things, music synthesizers, or whatever, and it was just great meeting all of these weird, cool, geeky people that made all that happen. Though it was not a requirement, we were all encouraged to eat lunch together every Friday. Over lunch, the lab coordinator, by now my adviser, would every now and then ask a philosophical question and not say much, encouraging people to discuss ideas.*

Exploring forms of communal living after his junior year of college led him to experiment as well with alternative forms of political organization as he started traveling across the United States in a Volkswagen van. *I explored notions of anarchy in which there were no permanent leaders, self-organizing to make cool things happen, and supporting each other to become better people. From there I found out about the Rainbow Gatherings from travelers who stayed in our collective houses, which were always open for guests, travelers, and all sorts of incredibly cool people.* Driven by his yearning for community, he became deeply inspired by the promise of the Rainbow Gatherings and the anarchist literature of the late sixties. He read the Situationists, Chomsky on his political takes, Jerry Mander on a social side, Alan Watts on a more spiritual side, and decades later learned about Hakim Bey and the concept of TAZ (Temporary Autonomous Zone), which he applied retrospectively to the interpretation of his experiences. *The Rainbow Gatherings took place in the woods, two miles away from any road, traditionally where there was nothing around, just the woods, and all of the sudden, there are thousands of humans, and within three weeks, there is a huge city out of nothing—hot and cold running water, ovens and places to shit that were designed to be healthy and ecological, without poisoning ourselves, health care, and all sorts of workshops—and within three weeks after the one-week event was over, volunteers made the place cleaner than when we were there.*

Another transformative encounter through the experience of the Gatherings was key: meeting its queer contingent and, in particular, the Radical Faeries. The main proponent of the Faeries, gay rights activist Harry

Hay, served as a big inspiration. *It is not that there is something wrong with me because I am this way, but that everyone, including gay people, women, and black people, introverted people, geeks, and Asperger's people, everyone has a place in society if we accept each other for who we are. Harry Hay started Radical Faeries because he thought gay people are unique and have something unique to offer society.* Communal experiences such as these taught Mitch to break out of his isolated, depressed world and to experiment commoning through electronics and through sexual and political identities. *By that time, I had quit all of my addictions and had managed to learn about a life that was not only about total depression.* Community was powerfully identified as an antidote to his feelings of inadequacy. The more he experienced shared spaces for electronics and exploration of sexuality, the farther he distanced himself from the blues. *Throughout my times at the Gatherings, I learned about various groups and subgroups within anarchism, how they all grew organically. No one really seemed to understand all aspects of it, but it was just this perfect whole. It did not mean it was just utopia; there were conflicts all the time because community is hard work. All relationships are hard work.* His last experience with communal living involved forming a collective intentional community in rural Tennessee called Idyll Dandy Arts (IDA), which still exists and presents itself as a "queer land space and educational project."[1] The collective experience at IDA got out of hand at some point, becoming unbearable to the point of physical threats, suicide attempts, and suicide among some of its members. It had become, in Mitch's words, an "evil cult," so he felt he had to part ways.

While still an undergrad, Mitch found his first computer job in 1977 at a software house in Champaign, Illinois, that created the first flight simulator for home computers. His job was to create a game using the same 3-D graphics engine used for a flight simulator. It was a temporary summer job, but he was exposed to the inner workings of the tech industry and came to realize some of the underlying connections between tech development and politics. He was surrounded by "technical wizards," who were "cool geeks" to him, but his satisfaction with the company did not last for long. *The military took the game that I developed, for which I dedicated three months of my life, and used it to make killer helicopter training simulators. They came and asked the company to modify it for their purposes, and the company said yes! So I quit, since I did not want to use my time and energy to harm other people.*

Soon after leaving graduate school in the mid-1980s, Mitch worked as an assistant professor at a local community college near the University of Illinois. He moved to Boston after two years of teaching and secured a steady job at a small company that made interactive museum exhibits using microcontrollers. Even though it was considered a well-paid, stable job by his peers and close relatives, he decided to quit after a year. People accused him of being irresponsible for leaving such a good job. He had settled in to his rent-controlled apartment after putting a lot of effort into remodeling it to be as he wanted it. From his apartment in a pleasant neighborhood, he had only a few blocks' walk to his workplace. But all the niceties did not prevent him from quitting. He found Boston culture too conservative, too judgmental about alternative forms of sexuality, too cold.

Before moving to the middle of town, the museum exhibition company had been located in the central hub of Route 128, the East Coast counterpart to Silicon Valley, feeding off the applied-tech contingent from several research universities in the region. His job had a typical startup organization. Internal infighting led him to quickly feel out of place. The top executives gathered proof to accuse one another of stealing from the company, leading to lawsuits and a very tense atmosphere. *Why am I working ten to twelve hours a day, busting my ass for these arbitrary deadlines, just for these assholes' profits? Other people working there were nice enough, but they, of course, were focusing on sales, so they were pushing* [our software and hardware projects] *to the military . . . where it was used to sell weapon systems. This is when I knew I had to quit. Again, I didn't want to use my time and energy to harm other people. This was actually the second time I made a choice for myself in my life. The first one was quitting television; the second one was quitting my job which I hated. And I hated living in Boston, so I moved away.*

After quitting his job in Boston, Mitch decided to travel without a fixed plan or destination, ending up in Alaska in his van. *It was the first time in my life I was truly out on my own. It was 1986. I was twenty-nine. While on the road, I realized that for the first time in my life I was happy. It was because I was making choices for myself. It was just me in my van. I could do whatever I wanted. It was all up to me what to choose to do. And amazing magical things happened. Seemingly magical things happen when you are doing things organically because it feels the best thing to do, rather than doing what you think people want you to do or what you should do.* From Alaska, he drove down the West Coast, stopping in Palo Alto, California, to visit friends working for local tech companies.

Based on a friend's contact, an artist from San Francisco who pioneered the work on computing and art, Mitch was hired to work with

MOS Technology 6502 microcontrollers for experimental applications around a project to create a visual program language. The company was called VPL Research (which stood for Visual Programming Language). It developed technology to create an immersive visual environment for its proposed programming language. Mitch worked on the electronics for a glove to move around and manipulate objects in 3-D space. The person who hired him and with whom he extensively discussed the possibilities of computer-generated immersive experiential environments was computer scientist and artist Jaron Lanier, who had coined the term *virtual reality* in 1987 to describe these immersive environments. *At VPL, we actually did not know we were working on virtual reality. We were working on input/ output devices for the visual programming language we were trying to create. But the I/O devices became the important part, especially when it became clear that there was no way we could create our proposed programming language with the technology we had available in 1986. So, we focused on the I/O devices, which is what became VR by 1987. I quit VPL after the military got a hold of one of our VR machines, one that I built myself, and used it as a World War III training simulator. Yet again, my work in tech was being used for harming others.*

Throughout this period, Mitch kept returning to the Rainbow Gatherings. He allowed himself to explore communal projects as a way of warding off depression. He eventually outgrew the experiences with the Gatherings and the Faeries. *I was progressing in his life with no patience for certain practices I felt OK with back in the 1980s and 1990s. In the beginning, I was very depressed, and I wanted to fix myself, because I was broken, or so I believed. But as I learned more, I realized we are all broken! Or, rather, it is OK. These aspects of ourselves we see as broken are just part of who we are. None of us are broken, you know. We are who we are. Either we see we are all broken—which isn't very useful—or we see that it does not matter. We can live and learn and grow and continually improve—while being who we are.* He recalls his parents did not know how to deal his homosexuality. They were not harshly homophobic but not entirely free from a subtle homophobic disposition either. *I grew up with a very prudish family, liberal socially in the abstract but not in practice. Me being gay, growing up in a homophobic society and them being clueless in how to support me while I was being beaten up every day for all different kinds of reasons. At school, people would beat me up and abuse me, physically and emotionally [sigh], so yeah, they were clueless about that.*

After working in the Bay Area for several years as a consultant, Mitch started a company called 3ware in 1997 with a friend to develop disk array storage controllers. The company did financially well for a few years, but left him seriously overworked, draining all his energy and greatly exacerbating his depression. *After the VCs [venture capitalists] took over [3ware], they tried to sell our RAID controllers to the NSA, so they could record all people's digital information to spy on all of us. I quit that job before that happened. It was such a horrible place after a few years. What was I doing? Everything I did was helping make the world a worse place.* There was no shared definition of Open Hardware in the late 1990s,[2] but 3ware was distributing the source code of their device drivers when the practice was very uncommon in Silicon Valley.

After leaving the company, Mitch worked as a consultant and saved enough money to support himself for a year without employment. He started to search for a project to invest his energies toward something he felt could bring out positive uses of technology. He volunteered for several collectives in San Francisco, including fixing old computers for nonprofit organizations and helping HIV-positive people through a telephone hotline. After heavily pondering the negative and positive aspects of technological development, an idea was suggested to him in a dream. It involved a magical device for masturbation that would provide people with immense physical pleasure. Reflecting upon his dream, Mitch found it really weird at first, but then he realized he could build devices with the same purposive action. *You could entice people into doing what they love with the device, except that the device itself was not really necessary. It was basically a way to trick people into doing things they love that they would not otherwise have given themselves permission to do.* The idea for his Brain Machine meditation glasses, an Open Source project to help people meditate, was inspired by this revelatory dream. *When I woke up from the dream, I thought, you know, that technology, the masturbator, was totally unnecessary—it merely gave permission, through the device, to do what people really want to do anyway but do not give themselves permission to do for whatever reason. It can trick people into doing what they love. So, I thought, "I could make a masturbator, but that is not really where I am at, so . . . what can I do?" And I realized that my idea for the meditation glasses was such a project. People clearly do not need electronics to meditate, but if it made it fun to try, then maybe it would give people permission to try it. What made it fun was that it used blinking lights at brainwave frequencies, an enjoyable side effect of which was hallucinating beautiful colors and patterns while meditating.*

Another of his projects of great influence was TV-B-Gone. It helped Mitch transform his TV addiction into hardware that was public in respect to its design, educational in its documentation, and international in its scope, as several hackerspaces wholeheartedly embraced the device as a must-have for the activist toolkit. He came up with the idea for the project in 1993. During a meeting with friends in a restaurant, he noticed how glued he and everyone was to the TV instead of paying attention to each other. *We started to talk about TV and its effects, and I thought, "Wouldn't it be great if we could turn these fucking things off everywhere we go?" And I realized I could. And one of his friends instantly came up with the name: TV-B-Gone. After I came up with the idea of the meditation glasses, I realized that TV-B-Gone was another one of those things. You do not need a TV-B-Gone to turn off TVs, but if you put this thing out into the world, it gives people the notion that they can turn TVs off, especially in public places, and it helps spread the word that turning TVs off is fun—and also mischievous, but that is the part that makes it fun. Then, you may find that life is actually better when TVs are off.* Developing the TV-B-Gone remote control literally meant hacking his addiction and helping other people deal with theirs. More generally, it also meant a form of communication through rebellion and collective action, refusing mainstream media culture wherever one goes, wherever there is a TV to be turned off.

Hacker conferences became key spaces for transformative encounters later in Mitch's trajectory. Maker Faire, a large conference for inventors, artists, and electronics hobbyists in the San Francisco Bay Area (but also elsewhere in the United States, Europe, and China), was one of the first events where Mitch connected with people who were working to bridge Free and Open Source hardware and software development. In 2006 he was invited to give a talk about TV-B-Gone at Hackers on Planet Earth (HOPE)—a major conference organized in New York City by the community around the *2600 Hacker Quarterly*. His experiences there convinced him to become more involved. *It was not until I started going to hackerspaces and hacker conferences that I felt totally at home in a group. With the Radical Faeries, I did initially, but I was still terrified of the various aspects of Faerie culture and its overt sexuality, which I loved, but it was scary for me at the same time.* After his first talk at HOPE, he was invited to attend his first Chaos Communication Congress, a mythical meeting ground for the European hacker scene, organized by the Chaos Computer Club (CCC). As an invited speaker, Mitch gave a talk on the TV-B-Gone project, which was received with great interest but also with heavy questioning. *I went the normal route: I am an inventor and my brother is a patent attorney, so I patented TV-B-Gone. That is what people did.*

Mitch took to heart the criticism he received from CCC hackers over his patent application. *When I was in Germany, I gave a talk based on the one*

I gave at HOPE about TV-B-Gone, TV and its effects, which resonated with at-
tendees, yet lots of people were asking me, "I noticed on the packaging, it says
Patent Pending. What is that about?" He contemplated the issue and came to
the realization it would be beneficial for him to go fully Open Source with
his projects. *I began again to think when I got home. You know, this has been*
kind of an Open Source project. So many people emailed me asking questions,
which I gladly answered, sharing whatever information about the project that
would help others. And many people emailed back with improvements in return.
People from all over the world were helping and publicizing my project for free. I
would then share their improvements with others. Why would I want to prevent
this exchange by closing the project to others? The more people make my project
and improve upon it, the more people will make things that turn TVs off, which is
what I really want from my project. So I just made it Open Source. More and more
people started helping as a result of that.

When questioned about his sense of ownership and his assessment
of the experience of going Open Source with his hardware creations,
Mitch pondered the existing constraints for an economy of common but
"intangible goods": *I tend to give all of my ideas freely as much as possible. I*
like the idea of Open Source licensing because people can use them for whatever
they want, but they have to share them with the same license and give me credit.
People know where the ideas come from, ideas that come from you, or the ones
that are unique to me. It is based on all of these ideas from other cool people. So, it
proliferates, and now people know where to look for more ideas from me and from
other people, and if they use stuff that gets shared back with me, I benefit from
it and everybody benefits. The only time there is trouble is when people violate
this agreement. In a powerful symbolic way, the gift of an Free and Open
Source project embodies the person of the giver: the provenance is person-
ified and not lost, so there is a channel of returning gifts.

At the 2007 Chaos Communication Camp in Berlin, Mitch attended
a preview for a talk on the best organizational practices for hacker clubs
and laboratories. This talk was entitled "Hackerspace Design Patterns"
and became a highly influential blueprint for the creation of hackerspaces
worldwide. It was particularly inspiring for Mitch, who, with a group of
hackers and artists from San Francisco, went on to create Noisebridge, one
year after his return from the Chaos Computer Camp. Meetings for the
creation of a hackerspace in San Francisco started, in fact, in the summer
of 2007 during the Camp itself. Among hackerspaces that were created in
cities around the world during this period, Noisebridge was particularly

known for its orientation toward autonomous politics and radical openness for all sorts of activities, including hardware and software workshops,
parties, political art, and autonomist projects.

As far as the origin story goes, the popularization of the concept of hackerspaces was a byproduct of an European tour by North American hackers
in 2007, who started, upon their return, to organize community spaces in
New York and Washington, DC, and a year later in San Francisco. In this
formative period, an international network of hackerspaces—initially limited to the Euro-American context but soon expanded to various countries
in the global North and South—took shape based on existing circuits connecting European and North American hacker collectives that stretched
all the way back to the first gatherings in the mid-to-late 1980s, such as
the historic Galactic Hacker Party. This technopolitical phenomenon represented, overall, a third wave of independent computing and electronics
associations, following the creation of electronics hobbyist clubs in the
1960s and 1970s, the exclusive computer clubs in the 1980s and 1990s still
operating under the guise of closed computing, and the experience of temporary autonomist spaces (so-called hacklabs) providing access and training for activists in the context of the alterglobalization movement in the
late 1990s and early 2000s.[3]

Mitch and other members of Noisebridge self-identify with the practice of autonomist politics, but what brought them together to form the
space was not political activism per se but the need for a space for hacking
with a common orientation toward the reinvention of technology. *I am an
anarchist more in the sense of liberty, not in the sense of a religion that you must
do things in a certain way otherwise you are wrong... and without the notion that
if you proceed this way, it leads to utopia. This is not the planet for utopia. I don't
know where that planet is. But this is my favorite planet, and this is the one I want
to help. But in order to help, I have to help as many people as possible to live lives
they love as they see it, not as I see it. I truly want people to do what they love. So
I go around helping people to form hackerspaces.* The term *hacking* itself has a
more contingent meaning for Mitch, one that is more attuned to its contemporary usage at hackerspaces, rather different from previous historical
definitions: *Hackers are people who do what they do because they love it. Whenever they are enthusiastic about something, they learn as much as they can, they
improve upon it, and they share with the world.*

Mitch collaborated actively with the transposition of Free and Open
Source software development practices to the context of hardware, result-

ing in very important interventions in the context of common electronics of the past decade: the proliferation of Open Hardware projects, associations such as the Open Source Hardware Association (OSHWA), and events such as Maker Faire and the Open Source Hardware Summit. Mitch's involvement was organic to the extent that, at the Maker Faire 2011 in New York, he was awarded the Maker Hero Award for his "outstanding contribution to the cause of maker-related education and/or open access to technology."[4] But the honeymoon of the nascent maker community did not last for long. O'Reilly Media, the former publisher of *Make:* magazine and former organizer of Maker Faire, announced in 2011 the acceptance of a $10 million Defense Advanced Research Projects Agency (DARPA) grant to fund one thousand makerspaces in US public high schools. In response to the creation of this strong tie involving the nascent Open Hardware community with DARPA, Mitch announced online that he had decided to part ways with Maker Faire but looked forward to work with it again whenever it no longer accepted military funding. The US Department of Defense sponsored the DARPA Mentor program with a budget of $3.2 billion for its first year. Mitch's statement ignited a huge controversy among members of hackerspaces and makerspaces worldwide. Soon after, he was invited to publish an opinion piece for the *Journal of Peer Production*, "Hacking at the Crossroad: US Military Funding of Hackerspaces,"[5] which would become very influential. *The tools and toys we work on and play with are very powerful. . . . Without taking responsibility for the technology we create, because of our cultural and economic context, our technologies probably will be used to feed this industry, and feed its military purposes. And all merely for profit's sake. Clearly, this bothers me. Clearly, this does not bother everyone. But, since it does bother me, I have made choices that, to the best of my ability, minimize the help I give to the military, regardless of possible profit.* After the publication of his piece, it became practically impossible to be indifferent to the debate and not be asked on which side of the controversy one stood when it came to the relationship between community spaces, governments, and companies.

Mitch has continued to support hackerspaces and Open Hardware projects to this day, devoting his time to giving talks and workshops in many community spaces and hacker conferences around the world. He moved out of San Francisco in 2017 due to the substantial rise in social inequality in the city, relocating to Berlin, where he has found community among technologists, activists, and artists.

THREE

Chaihuo

Between Gifts and Commodities

It was a mild late-October evening when Mitch Altman finished one of his signature talks on the importance of "doing what one loves with technology" at Chaihuo, one of the most cosmopolitan community spaces of Shenzhen. Promptly, someone in the audience raised a hand and asked: "This is something you can do in America, but China is very different. How can we make money with Open Source here in China?" At these meetings, a similar question nearly always came up. Mitch then explained his own experience of making a living with his electronic inventions. In the audience, we had dedicated members of the Shenzhen Do-It-Yourself (SZDIY) collective who would later confide how embarrassed they were to have an "unknown person" in the crowd ask a question of this nature: "He does not understand what Open Hardware means." Different perceptions of this unusual economy of gifts among strangers were, in fact, always in question—not that gift giving was not key for comprehending social ties in mainland China,[1] but it took rather different forms than it did in the Euro-American circuits of open technology. Following Mitch's response, one of the founders of SZDIY passed throughout the audience a Tupperware box with a surprise inside: a clone of Mitch's TV-B-Gone. It had the same prototypical look: a bare circuit board with visible through-hole components soldered onto it, an array of exposed infrared LEDs, and some silkscreened text. Mitch was caught by surprise but smilingly declared in front of the attentive, young crowd of Chaihuo habitués that he was very thankful for the gift—in effect, a counter-gift that returned to one of the initiators of a series of contributions, animating various forms of circulation (of material and immaterial gifts) within an extensive common circuitry.

Itinerant hackers from various hackerspaces made yearly "hacker tours" to China to present and exchange projects, purchase electronics parts and tools, and get contacts to manufacture Open Hardware technologies. In Shenzhen, Chaihuo was one of the obligatory ritual stops. I first joined a hacker tour in 2013 while doing fieldwork in Southern China, where I was hosted by a couple of inventors from China and the United States who ran a small open technology business from the living room of their apartment in Bao'an, an industrial district on the periphery of Shenzen. My days were packed with visits to local companies and technologists working with open technologies, mind-expanding Buddhist restaurants, community events, and electronics markets. The best portion of my time was dedicated to informal meet-ups and electronics tinkering alongside SZDIY members.

Chaihuo represented a rather curious place in the hackerspace network where itinerant foreign technologists could meet local engineers, students, software developers, entrepreneurs, and designers. The space was conceived in the encounter of SZDIY members with foreign engineers. Among the original members are two Chinese entrepreneurs who operated a company in a small rental apartment, called Seeed Studio, which would eventually become one of the most influential Open Hardware manufacturers and distributors in the world. Chaihuo was first established in an office space in Nanshan district and later moved into the Overseas Chinese Town Loft (OCT Loft) in an upper-middle-class neighborhood run by big real-estate investors. The OCT Loft was an exclusive place: an upscale complex with art galleries, studios, boutiques, jazz bars, restaurants, and coffee shops. In its new iteration, Chaihuo was decorated to match the surrounding shops with modular IKEA furniture, ambient lighting, and lime green detail paint on the walls. The negotiation with OCT administrators involved having to explain that "makerspaces"—disciplined variants of hackerspaces—were related to the neighboring shops and galleries. "We had to convince [them] we were about art, that [Chaihuo] is close to the arts. You see, the neighborhood, they want to maintain the style. They don't just rent to those paying more," explained Seeed founder Eric Pan.

The first years of Chaihuo coincided with the formation of the SZDIY group in 2011. Their early collective experiments included common hackerspace activities, such as pair programming, brewing beer, teaching one another computer-aided design (CAD) for drawing circuit boards, and peer-mentoring with open technologies. Their common space was one of conviviality, where open technologies that could be studied, shared, modified, and redistributed, according to philosopher André Gorz's definition, were diametrically opposed to "locking technologies."[2] Hardware hacking sessions, for example, involved shared meals and birthday celebrations, as well as frequent visits of hackers from other spaces abroad. The motivation to start SZDIY grew after two co-founders attended a talk by Richard Stallman at Shenzhen University in 2009. From this encounter, they started to look for Free Software enthusiasts to form the Shenzhen Linux Users Group (SZLUG). In their weekly meetings, tech enthusiasts gathered to demonstrate software or hardware projects, conducted workshops, and prepared prototypes with 3-D-printed parts. Each "demo" instantiated locally gifts that circulated as open technical objects, relying on extensive online and offline circuits.

The CAD software KiCad, for example, was made common by a French electrical engineering professor and later revamped by eager staff engineers at the European Organization for Nuclear Research (CERN) to become a very popular and well-maintained community project.[3] Its sociotechnical extensions would reach as far as Chaihuo to become the tool of choice for community workshops. 3-D printing was placed in common by a group of enthusiasts around the Replicating Rapid Prototyper (RepRap) project, started in England at the University of Bath. The goal was to create a platform for making 3-D printers that could replicate their own parts, while promoting, in the words of its inventor, engineer Adrian Bowyer (2004), in his manifesto "Wealth without Money," "the revolutionary ownership, by the proletariat, of the means of production."[4] This was all made possible when the temporary commercial monopoly on the technique of fusion deposition modeling (FDM) legally expired, leading to an exponential growth in China and abroad of 3-D printing projects and, subsequently, startup companies that captured, in science and technology scholar Johan Söderberg's evaluation, the transformative potential of the common for the consumer market.[5] Beyond hardware and design tools for hardware, SZDIY members also congregated around programming environments, libraries, and copious documentation that linked them with old-school hacker groups, such as the Lisp community. These examples in software and hardware represented just a tiny sample of the key infrastructural technologies (from other cosmopolitan circuits) for the common instantiated at Chaihuo (figure 3.1).

Every community space is punctuated by its own rhythms, affordances, challenges, governance practices, and tensions that condition practices of exchange. Just like Noisebridge, Chaihuo had its very own particularities. During regular workweek hours, the space stayed mostly quiet and calm. I often came and shared the main table with young engineers working on their own commercial projects: a wearable Bluetooth speaker, a robotic arm, and a "social-networked" bracelet. I immersed myself regularly in tinkering with microelectronics tools to identify problems but also participated in the discussion of solutions with other members at regular Thursday meetings. It was common for me to arrive at the space and find a recent engineering graduate working quietly on a project, preparing printed circuit boards on the spot. After finishing the 3-D-printed casing for a device, I might find him hunched over his circuit boards with a toothpick at hand, applying soldering paste patiently on very small contact

FIGURE 3.1   Regular SZDIY meeting at Chaihuo

pads to place surface-mount devices (SMD) one by one—a task otherwise performed by automated computer numerical control (CNC) machines that hackers insisted on learning to do in an attempt at re-enskillment *after* industrial deskillment. The labor-intensive task of "populating" the boards would, then, lead to the actual soldering of very small components. For this final step, the populated printed circuit boards (PCBs) would be placed in a reflow oven—one of the few fabrication tools of Chaihuo at that point, other than a 3-D printer, a few soldering stations, a deactivated laser cutter, and tools for basic electronics diagnostics. Chaihuo had bins and drawers where one could find prototyping parts, boards, and tools for electronics projects, similar to Noisebridge's "hacker shelves," but most fabrication work did not take place on-site. Shenzhen was considered a sprawling workshop for fabrication and microelectronics for projects large and small, so there was no shortage of places to get help. Where we found the bins with tools and parts, we also found a napping pod—frequently used by overworked tinkerers. The hard-won legacy of delimited "work hours" that the social movements left to us was not really a thing for the young engineers at Chaihuo: everyone worked constantly, except during traditional holidays, when every business came to a complete halt.

It was common to find foreign entrepreneurs from North America and Western Europe looking for talented engineers at Chaihuo. Many foreign projects involved soulless consumer products, such as an Internet of Things–based "dog entertainment" machine, meant for crowdfunding on Kickstarter, a popular website for organizing collective financing campaigns. SZDIY members were not particularly excited about or available for these head-hunting sprees. Most of them had regular jobs in local or transnational IT companies, so it was not uncommon for the space to be devoid of the convivial effervescence of public workshops during the day. The peak hours for most hackerspaces coincided with open-house events during evenings and weekends, and so it was for Chaihuo. Outside these special hours, only paying members could be found working on projects for domestic and foreign markets.

In organizational terms, Chaihuo was managed by community members who traded their time and service in exchange for waiving membership fees and permission to use the space for their startup projects. Contrary to one of the key independence "hackerspace design patterns," Chaihuo was supported primarily not from donations and membership fees, even though they were requested, but rather from the direct sponsorship of Seeed Studio. The community was very small in its early years; there were less than two dozen active members. Its co-founder, Eric Pan, insisted on defining Chaihuo as a "makerspace" when he first discussed the project with the SZDIY collective. The loanword *hacker* (黑客) is part of everyday parlance, but it is also used in the Chinese news to refer to criminal activities. Despite their recognition of Euro-American hacker-origin narratives, most Chaihuo members—with the exception of the SZDIY collective—would prefer the term *maker* to signal their strong entrepreneurial orientation in a moral landscape that anthropologist Yunxiang Yan has characterized as aggressively individualized.[6] Young engineers were well-aligned with the promises of "making" as a community-driven form of "innovation," but more importantly as a means for "prosperity." In contrast with Noisebridge, Chaihuo was not open to nonmembers except for open-night events when an acrylic laser-cut box would be passed around for donations.

What distinguished Chaihuo from other spaces was its proximity to the electronics market of Huaqiangbei. In crammed booths of this gigantic sacred site for electronics, we could find sales representatives of various factories with samples of electronic components and tools of various kinds, LED shops, computer and cellphone repair booths, and retail shops of all sorts, where young shopkeepers, who were often women, could be found soldering circuit boards and fixing laptops with incredible dexterity. Being at Chaihuo was very convenient for SZDIY members to go on regular trips to "the jungle," as the market was called, to hunt for electronics parts in the frantic maze of tight and crowded stands overflowing with parts, cables, and tools. With the overpowering aural and visual stimuli, it was easy to feel disoriented. It was also fairly easy to find what one was not looking for, as nearly everything could be obtained for free as a sample or for a fee—from security-protection cracking of chips through "depackaging" to difficult hardware repairs (of ball-grid array soldered parts in cellphones, video-game consoles, and personal computers) to creatively shaped portable phones and older models of electronics testing equipment (shipped in containers of e-waste to Hong Kong and then to Shenzhen from "first-world" sources, as they are locally known). Huaqiangbei animated an underground world of electronics repair that had no parallel: small computer shops had access to the engineering drawings of popular computer parts and circuit schematics that were most likely leaked from the numerous OEM (original equipment manufacturer) factories in the region. One could find repair help and replacement parts for virtually anything. The market felt like an enormous, sleepless high-tech bazaar with a strange mix of grandiose modern architecture, human precarity, and brightly lit LED panels.

A Hacker-in-Residence

One of my regular partners in excursions to "the jungle," nicknamed Atommann, made himself known as a hacker due to his profound passion for science and technology. He was not only one of the founders of SZDIY but also a fixture at Chaihuo since its inception. He could always be found at the space helping someone with a technical problem or presenting one of his hardware experiments. Once, he helped me resolder a micro-USB connector that had snapped off from a microcontroller board (figure 3.2), a practical lesson that encouraged me to take on much more challenging soldering jobs. In addition to electronics and computing, physics was one of Atommann's key areas of self-education, something that came to him through an unlikely event. "I found a book in the rice paddy field at home," he once said, "and [took] a long bamboo pole to fish it out. It was a

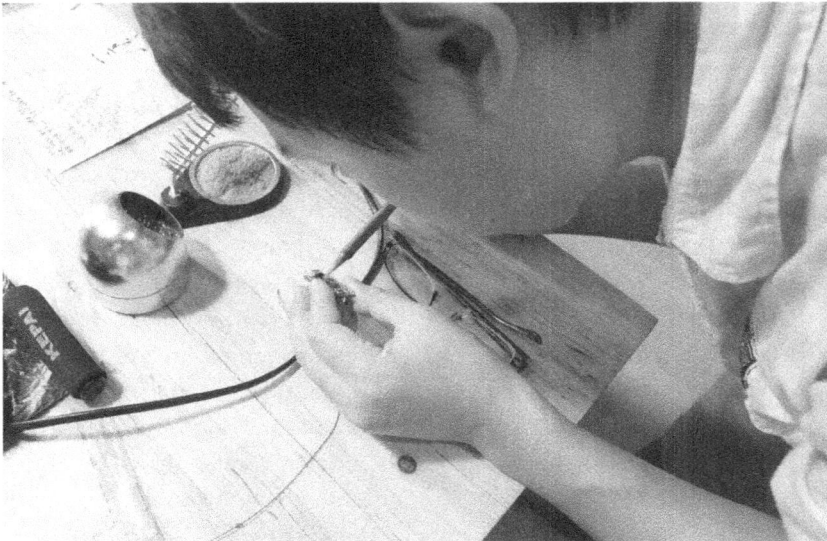

FIGURE 3.2 Atommann fixes a broken micro-USB connector on a circuit board

junior high school physics book. I read the cover and it mentioned electric motors, diesel engines, generators. . . . It was as if I had found a treasure. I dried it out and enjoyed it every day." He soon started experimenting with all sorts of materials and their physical properties to create wind turbines, motors, and waterwheels. Everything he could put his hands on, he would repurpose for his experiments. Only later in life, as a young adult, did computing become one of his obsessions—he saw it as a concretized marriage of physics and math. His deep appreciation for the field led him and his partner to name their two young daughters after celebrated computer scientists and mathematicians: Grace (Hopper) and Ada (Lovelace).

After working with industrial automation, Atommann found employment in an international hardware incubator, HAXLR8R, of foreign investors and hackerspace founders, from France and the United States respectively, in Huaqiangbei. When speaking about his training, he would proudly declare himself an autodidact like many other hackers worth their salt. His first contact with computers happened when he was seventeen, as he did not have access to digital technologies in his natal village in Sichuan province. At Chaihuo, I would often find him on his computer reading Wikipedia articles or technical books in English when he was not helping someone troubleshoot a software or hardware problem. Lack of formal training did not prevent him from advancing his projects, such as a "picture phone" that he designed and fabricated from scratch and whose prototype could be found on display in Chaihuo's storefront. Atommann had made it for his grandmother. As she aged and could no longer remember her relatives' phone numbers, all she had to do was click on a picture to be able to talk to them. He also worked on several other small projects of his own, such as a temperature probe that reported the temperature in Morse code (made as a joke) and a "breadboard" computer made from discrete transistor chips (made for self-educational purposes).

During an afternoon like any other, two Chaihuo members and I took the Shenzhen metro from OCT to the residential area of Nanshan in search of an affordable shop where we could run design files on a laser cutter. On our way, we stopped for tea at a small shop on a tree-lined street with swirling branches hanging over the sidewalk, relieving some of the humid summer heat. At a certain point in our stroll, Atommann pointed out a symbol carved into the pavement: "Look, a diode!" He placed his foot at the top end of the letter F in the combination KF to form the symbol of a diode used in circuit schematics. I looked at it and visualized the symbol only after trying to dissociate mentally from their conventional sounds. The person accompanying us, an economics major and hobbyist computer programmer, commented in a dismissive tone: "It looks like a K and an F to me." We never managed to cut our design files, but this rather mundane occurrence reminded me of a shift in my own perception after attending workshops on "lock picking" at hackerspaces. I could not help but pay attention to locks and wonder what mechanism they hid and, most importantly, if I could open them. At this point I caught myself thinking, "Why am I obsessing with locks?" It was for the same reason Atommann was obsessing about electronics components, I realized. He would spend a substantial part of his day going through open design files (on KiCad) of various projects in his capacity as hacker-in-residence for the hardware incubator. The symbols of his daily technical practice would inform his perception in important ways in this telling instance of phenomenological modification. After going through an informal cultivation in common circuits, it became second nature for him to perceive a diode where others would see only alphabet letters.

Making Automata, Managing Chaihuo

In terms of cosmopolitan cultivation, Chaihuo was animated by young engineers who were integral to specific circuits: they had either participated directly in community spaces abroad or entered the fast circuits of global circulation by following and contributing to Free and Open Source projects and conferences. One of the most influential Chaihuo managers, Hao Zhang, had his debut at Noisebridge, where he worked alongside other hackers on a robot that operated a wheelchair. His project was called Dorabot after the Japanese manga Doraemon that featured a time-traveling robotic cat whose mission was to help improve the life of his human companion by using fun and loopy gadgets. Dorabot was initially an open platform for making robots, but eventually, it became an established company dedicated to industrial automation. Dorabot was an organic part of Chaihuo's hardware workshops after Hao returned from his hackerspace pilgrimage, bringing back with him everything he tacitly and actively learned abroad.

Hao's trajectory has many parallels with other Chaihuo members, with one key distinguishing factor: he was one of the first to be exposed to the early development of hackerspaces outside mainland China. Growing up in a well-to-do family in the province of Fujian in the 1990s, Hao had early access to digital technologies arriving through extra-official channels. He got started with computers in the form factor of game consoles in elementary school, like many hackers of his generation. He told me: "There was a very popular knock-off of the Family Computer (Famicom). It looked exactly like the FC, but we did not have the FC; it [was] not legal to sell it in China. I was in primary school when I got one. It was an FC with a keyboard. The keyboard was designed for learning. I remember clearly my father buying [the educational game] Logo. It was very expensive. Almost as expensive as the machine itself." His earliest experiences with programming consisted of modifying scenarios for the popular game Mario Bros. He became too obsessed with it, so eventually his parents decided to forbid its use. He had to go to his neighbor's house to play.

Hao learned English early at the insistence of his parents, Party members who were interested in providing the best education for the twenty-first century he could get. When he was in fourth grade, his father purchased him his first personal computer, a Compaq, which cost a fortune at the time. "My father was very open-minded," he told me. "Everybody was saying that there were three [skills] that [the Chinese] must have in the twenty-first century: computers, English, and driving." Hao started learning BASIC and then Pascal, which led him to join programming competitions. His interest, however, was in learning C instead of Pascal. "In every textbook, they would say that Pascal uses English, natural language for control symbols, so it is easier for you. But we are Chinese, we do not need that [*laughing*]!" He started to get into robotics in college through Robocon and felt compelled more and more to start his own project: "It was like creating a new form of life, except that it was not." He was a quick thinker and an articulate speaker, so he was invited to join the debate team in college. He excelled at debate and was eventually invited to coach the university team. The topics of debate, however, did not touch upon politics at all: "We have this idea that if you talk about politics in public, you might get in trouble. We do not say bad things about the Party in public." Hao enrolled in several science and technology courses, but they were not that compelling to him. His university held electronics design contests, mostly dedicated to improving the efficiency and accuracy of testing equipment. "I was not interested in how much current goes through a lamp. What is the efficiency of your system? I was not interested in that," he said. "I was interested in what it could do for you, in building something novel. That has always been my principle, my way of thinking or way of doing things. That explains everything I ever did."

Noisebridge created a lasting impression on Hao—its openness, solidarity, nonconformity, and lust for collective experimentation. His experiences in the Bay Area sparked his interest in playing a more central role in his local community and also in starting his own robotics company based on Dorabot, which he ran from Chaihuo when he was not offering public workshops: "Not sure how familiar you are with Mitch [Altman] . . . but he had socialist experiences [with counterculture]. In socialist places or communities, theoretically, people are supposed to treat each other very well, like they did at Noisebridge. Everybody can join; everything is pretty much free. You can do pretty much anything you want as long as you are nice to the place and to each other. It is like a utopia . . . but . . . as we know, none of the utopias work. Noisebridge [was] definitely amazing. If it wasn't for [its] level of openness, being as broke as I was, I would not have had the chance to build a prototype of Dorabot."

Hao did the prescribed pilgrimage of hackerspaces in United States and Europe. He visited MITERS (at MIT), metalab (in Vienna), NYCResistor (in New York), and the Chaos Computer Camp (in Germany), but Noisebridge was the place where he was best received. Before leaving for his international tour, he became involved in the creation of Beijing Makerspace, but it eventually became more of a co-working space for startups. "I first heard of Noisebridge at a dinner when Mitch gave me a key. This level of confidence in people is amazing." Hao was hosted by Noisebridge members during his visit to San Francisco and slowly but steadily became very involved in the sociotechnical life of the space. In his experience, people were extremely accommodating. They provided access to equipment, resources, and parts that he needed, in additional to overall support for his work: "When I lost my glasses at Toorcamp [a hacker gathering on the West Coast of the United States in the mold of the Chaos Computer Camp], I could not afford to get new ones. I could not afford the Noisebridge membership. First, Jake [Spaz] tried to access public services to get me another pair. He knows a lot about these things [how to live austerely]. Most of these services required you to have a resident certification. Even if I were a San Francisco [homeless] person, I would have some sort of card. I needed to go back to my project, and I needed glasses. There was a TED Talk we found about making a cheap pair of glasses for poor kids. So Jake just built me a pair." Communing with people in the hackerspace community was fundamental for his ability to circulate in the network with minimal resources and minimal friction: having common experiences in hardware hacking circuits opened doors.

Hao's experiences at Noisebridge furnished him with an opportunity to expand his understanding of hacking as well, including "food hacking. [Noisebridge members] used plant material for making meat substitutes. This was pretty amazing for the United States, but was common practice here in China. When people refer to this kind of food here in China, they call it 'fake meat.' In the United States, they call it 'vegan' or something like that. This was the first time I realized, 'OK, we can hack food.' [At Noisebridge], they had this sewing area, so we could hack clothing. They had this dark room, so we could hack photography. I have never thought [these activities] could be thought of as hacking. Basically, tinkering with everything you can find that interests you, that is hacking."

From Gifts to Commodities (and Back)

To understand the importance that Chaihuo played in the *spatialization* of hacking, it is important to examine in more detail the differential roles of its core constituents: SZDIY and Seeed Studio. They are highly relevant to common circuits because they describe distinct trajectories in building interfaces with hackerdom. While SZDIY carved out a space for independent tinkering and collective self-directed learning with open projects, Seeed followed an alternative business model grounded in experimental projects not only from SZDIY but also from others, such as Dorabot. They modulated network signals, in other words, with a different moral circuitry, but within the same political economy. Different types of circuitry represent what philosopher Gilbert Simondon called an "associated milieu," a fundamental, associated natural, technical, and social context for the process of individuation of technical objects.[7] Seeed was integral to the emergent entrepreneurial milieu of Shenzhen. The SZDIY collective, on the contrary, was primarily invested in promoting a space of conviviality around hacking as a hands-on practice, so it eventually grew apart from Chaihuo, establishing its own independent space in 2014.

In the span of two fast-paced years since its inception, Seeed gained international recognition and, subsequently, national notoriety. Its founder and CEO, Eric Pan, made the cover of the special 2013 issue of *Forbes China* on the "30 CEOs under 30," thus validating Open Hardware as a profitable business model in the eyes of investors. The company also attracted the attention of high-ranking politicians, such as the premier of China's State Council, Li Keqiang, who visited Chaihuo in a 2015 tour of the Guangdong province. The end of the story of this heavily mediatized and celebrated encounter was well-known, if not skeptically anticipated in the hackerspace community. Hundreds of "innovation spaces" were created in China between 2015 and 2017 as part of a governmental reform to update industrial production through the circuitry of "knowledge production," but were subsequently dismantled one by one for not fulfilling the (institutional) promise of "grassroots innovation." In her study of entrepreneurialism in postsocialist China, informatics scholar Silvia Lindtner examined the policy of mass innovation in terms of a process of neoliberal self-making.[8] The official projected growth of makerspaces in the mainland—as reported by national and foreign market analysts alike, such as Gardner and Tencent—predicted a few hundred in 2014 but several thousand in 2018, numbers that would greatly surpass those of existing community spaces worldwide. As SZDIY community members suspected at the time, this cheerful balloon of corporate and governmental hype would inevitably pop: the spaces that were funded by the government and created overnight had no community to support them whatsoever. They were left mostly empty and under-resourced as a result. This failure of politico-economic evaluation contained a very simple, yet hard-earned lesson: one cannot create community and conviviality by decree—a fairly common failure, we could say in retrospect, in the ability to create a space for commoning that independent groups, such as SZDIY and Noisebridge, had enabled in their own way and by their own means. The hackerspace design patterns document was prescient in this regard for anticipating the anti-pattern of "institutional sponsorship."

Beyond Chaihuo as a space for encounters across a wide range of technical and political differences, Seeed played the role of cosmopolitan hub as it placed local engineers into Euro-American circuits of hardware production, animating a circuitry with potential for non-alienated and convivial work. As they approached hegemonic circuits, Seeed employees started to learn about open technology development and licensing practices by attending conferences such the Open Source Hardware Summit and Maker Faire in the United States. Attribution, originality, and "proper copying"—studied in anthropology by Cori Hayden and Gustavo Lins Ribeiro[9]—assumed radically different moral connotations in this context, creating new conditions for Seeed to convert gifts into commodities (and vice versa): "What [people in China] are trying to do," suggested Eric, "is to merge with the global Open Source world, but it is not so widespread. In China, the copyright system is not working very well, so only people in Open Source will do Open Source. It is very easy to reverse engineer something here. One example is the case of mobile systems. Many factories in Shenzhen have been making mobile devices in the last five to ten years within their ecosystem, within their community. It is Open Source without a license—exchanging ideas, refining designs."

When describing his early professional experiences before Seeed, Eric recalled the encounter with the world of small-scale electronics manufacturing that changed his trajectory after leaving a prestigious internship at Intel: "After one year, it started to get boring because I could see ten years ahead. So I quit that job and started packing to leave for three months without a destination. Especially in the Chinese educational system, you are taught to go from A to B and find a solution to solve a problem. All of a sudden you need to find a destination that suits you. I went to Beijing, tried all kinds of opportunities, even working for an international trade company, and interviewed for all kinds of companies. I could not find a very good one, so I tried to do some startup work. There was no big plan, no big dreams, just to find something to survive on at first. I did some work on Taobao [online e-shop platform], did e-commerce, but that did not go well, until I encountered a new media exhibition in Beijing, and I saw how artists play with apps. I think it's a very interesting industry. It's very open and very communicative. I talked to people and that's sort of where Seeed started." Chaihuo followed the creation of Seeed, but not without the necessary grounding in the convivial circuitry of SZDIY.

Many technologists in the hackerspace network viewed the "impossibility" of intellectual property (IP) enforcement as generally a "good thing," since it allowed Chinese engineers to participate in a common, much less constraining than the litigious environment in Euro-American countries with strong corporate IP control. From this external and rather romantic viewpoint, China embodied the "utopia of a world without IP" for creating alternative exchange practices similar to Open Source–based moral economies (but without the legal device of open licenses). On a sociological register, exchange practices hinged upon the moral economy of *guanxi* to create a well-sheltered in-group.[10] IP compliance among technologists became possible in China in part through Open Hardware collaborations involving a small number of Chinese hackers with ties to foreign projects. This process has been thematized by hardware hackers, design scholars, and practitioners—such as Andrew "Bunnie" Huang, Silvia Lindtner, and David Li respectively—in terms of a fundamental shift in southern China toward the revaluation of local manufacturing as a site of creativity, ingenuity, and self-reliance.[11]

In the anthropological debate about gift-market relations, one of the most fruitful questions is how gifts become commodities and how, conversely, commodities shift to the register of the gift while maintaining their capacity for reverse transformation and differential valuation.[12] This debate is particularly important for understanding what is at stake in the circuitry of exchange relations involving independent hackers (from SZDIY) and startup ventures (at Chaihuo) with companies (such as Seeed Studio). Since technical exchange is one of the main ingredients of hackerspace sociality, practices of commoning around open technologies have a tendency (not always positively realized) to create ties among strangers and compose larger and more distributed collectives. When short circuits happen (as they often do in community spaces and projects), they have to do with the fact that most entrepreneurs are not practicing hackers. Whereas the former are in the business of taking but rarely reciprocating, the latter are often the ones who carry the proverbial piano of building technical objects and sharing widely. For Seeed to convert Open Hardware projects into products, for example, its collaborative ties have to be created on the register of the gift, following a cycle of contributions and counter-contributions that is grounded in the hackerspaces network.

It all starts with an open technical object, followed by a cycle of research, redesign, sourcing, manufacturing, and redistribution with its beginning and endpoints at community spaces. In the context of what anthropologist Anna Tsing has called "supply chain capitalism," open technology projects are made available as gifts among strangers with the potential of being converted into commodities.[13] Intangible elements such as firmware code, documentation, and hardware design files are publicly released, but there are also numerous cases in which they are only partially or belatedly made available to prevent competitors from "rushing into the market" as a signal of a morally questionable shift of register. To complete the process of conversion of gifts into commodities, Seeed relies upon the network to provide technical expertise and creative input. And to accelerate the process, the company sponsors Chaihuo to serve as a storefront for Seeed products as well as for community projects and as a meeting point for local groups and itinerant technologists from "hacker tours." This whole set of prestations—to use Mauss's expression for obligatory, "total," or, better put, "general" exchange practices—animated not only workshops for hardware enthusiasts at numerous convivial places in the network but also a conventional industrial supply chain—the invisible side and constitutive "outside" of the moral circuitry that is composed of precarious workers that do not get to participate in cosmopolitan exchange circuits.

Seeed's entrance gate resembles the entryway of any other Chinese industrial enterprise, except for the products on display: products that were once open prototypes requested to be manufactured in China by foreign designers. One such project was a garden timer and controller called Open Sprinkler, the quintessential product of North American suburban life. This "open product" was created in partnership with Chris Anderson, pundit of the US-based maker movement, calling attention in its packaging to the absence of the omnipresent "Made in China," now replaced by the Seeed's new aspirational slogan "Innovate with China." At first sight, it suggested Euro-American hardware designers and engineers could innovate with China, having the resources of the manufacturing region as an associated milieu. It was rather the case that cosmopolitan circuits were supported by the material infrastructure of subcontractors that rely on exploited laborers, where the actual makers are not invited to participate in the adventure of "making." A relatively small number of technologists get to participate in the cosmopolitan circuits, whereas the biggest number of industrial makers on the factory floor are relegated to the alienated portion of the work. The study of the common always demands that we ask, Common for whom? Common of what and under what historical, technical, and socioeconomic conditions?

Seeed Studio's supply chain is instantiated under particular conditions that can be described by reconstructing the steps of open design, industrial production, and global distribution of an Open Hardware project: the Flying Stone Tiny (FST)-01, a Gnu Privacy Guard (GnuPG) USB token for data encryption created by Niibe "Gniibe" Yutaka (whose trajectory we discuss in chapter 6). While working for the GnuPG project with German hackers, Gniibe started to implement a random-number generator in a low-cost platform to create a cryptographic token as a substitute for Pretty Good Privacy (PGP) smart cards. In his first attempt, he used a microcon-

troller unit (MCU) from the AVR family, the same platform made popular
with Arduino boards in most, if not all, hackerspaces. The performance of
this MCU was too slow for computing cryptographic functions, so Gniibe
tried another chip, a very small 32-bit ARM computer manufactured by
STMicroelectronics and distributed in an educational product, the STM
Discovery Kit. He had the idea of repurposing the kit by literally hacking
the board in half and using just the portion with the STM32 chip for his
security device. After validating his very first prototype at home, Gniibe
started to search for a printed-circuit-board manufacturer and found
Seeed Studio online. He then submitted a "wish," using Seeed's website for
his project entitled Minimum STM32F103 USB Board. User-contributed
"wishes" at Seeed allowed the interest of internet users in open technology projects to be evaluated. It was an entry point for project ideas to be
tested for commercial viability by the company.

The FST-01 was not a usual Open Hardware project for important reasons. Gniibe used a very small computer that was not common for hackerspace projects at the time. He also designed his device using KiCad at
a time when it was not the most user-friendly tool. And, finally, he implemented the cryptographic functions plus a random-number generator
of a high degree of technical complexity, which was in heavy demand by
hackers and security professionals but still quite esoteric for most computer users. We must remember how much fuss this technology generated
after Edward Snowden's revelations of mass surveillance in 2013. One of
the journalists involved in the NSA leaks, Glen Greenwald, was very vocal
in his distaste for GnuPG as a technology that was very hard to use. That
critique, of course, did not take into consideration the solidarity behind
the tool, which took the effort of several hackers contributing to the digital commons.

In order to test his first prototype, Gniibe got in touch with Seeed to
request the fabrication of a small quantity of printed circuit boards. One
of the women engineers at Seeed took up his order and went over the requirements. Gniibe used Seeed's wiki to distribute documentation in
English about his project, which he copied verbatim from his personal
website, where most of the documentation was originally written in Japanese. The wiki page for the project was edited to provide links to public
repositories where he published all the necessary files for replication, including firmware code, schematic drawings, and other design files. It also
listed the bill of materials (BOM) with the name of online distributors

for more specific electronic parts, such as the MCU unit STM32F103. In the meantime, Gniibe started to put together his own USB tokens, using an improvised hot plate to solder tiny surface-mount electronic components at home. He validated the hardware fairly quickly and went on to order a "larger" batch (+100) from Seeed Studio's Propagate service, which became a distributor of the security token from that point on. Seeed described Propagate as a service for agile manufacturing for small-batch PCB, kitting, and assembly. It required, before manufacturing, estimating the return on investment of any project. This estimate was prepared for the client after obtaining the BOM to be used, plus a test plan for quality assurance. Finally, public distribution ensued with documentation in English on the wiki with a link to the product to be bought at Bazaar, Seeed's e-shopping platform. Following all these steps, Gniibe's hardware project was submitted in July 2012 and manufactured in August 2012. By the end of September of the same year, the production was done and the product FST-01: Tiny USB 32-bit computer was ready for distribution.

What is unusual about this case is the imbrication of common and privatized instantiations of open technologies as they travel from their contexts of invention, published as gifts, to end up as commodities—that is, from the sphere of activity around private electronics labs and hackerspaces to manufacturing locations and back. In order to sustain Open Hardware projects, relations must be created on the registers of the gift and the market, following a general pattern that starts from an existing open project and undergoes a full cycle of research, development, sourcing, manufacturing, and distribution. Once distributed, the project is open to reenter the cycle through a derivative version, just like the SZDIY version of Mitch's TV-B-Gone.

The feasibility of open projects through production, circulation, and usage is guaranteed by the possibility of shifting registers between gift and commodity relations. Open technologies can be a gift whose ties may and usually do extend to market relations, involving technical consulting under service contracts. Only in some commercial applications or appliances is Free Software embedded in a commodity. Open Hardware, conversely, is born as a gift carrying the likely potential for becoming a commodity and a service. Its intangible constitutive parts (firmware, documentation, and design files) are released publicly in digital form for wide distribution. Its tangible parts are held back to be exchanged and evaluated on the register of the market and within industrial settings. In order to sustain the commoditized aspect of Open Hardware, instances of gift giving are created with the distribution of intangibles as well as in face-to-face workshops in hackerspaces, where the novice comes to learn about electronics, receiving hands-on instruction through the means of an open technology kit. Hackerspaces, by their turn, create sociotechnical conditions for the conversion of gifts in commodities and vice versa: they bring together technologists to share information about their projects and animate the flow of gifts through pedagogical encounters, as well as, at times, entrepreneurs who are the main agents in the conversion of gifts into commodities in spaces that are more welcoming to them, such as Chaihuo.

Open Technology Factory Tours

On my first visit to Seeed's headquarters in the industrial district of Nanshan in Shenzhen, I was escorted by the chief of human resources (HR) and told that the company was "run as a big family." Her team frequently published an internal newsletter with employees' birthdays, movie recommendations, and discussions of group activities, regularly posted on the walls around the company. The HR department also organized group activities. The engineering team, for example, met weekly for soccer matches, without any woman-identified engineers, at a nearby soccer field, which I also joined. Entering Seeed's office, one passed through a meeting room facing a large open area. Aesthetically, the scene would fit right into Silicon Valley: there were no cubicles and everyone worked side by side in colorful surroundings. There was also no fixed division between teams as they were allocated by managers and rotated tasks. The most prestigious engineers were also habitués of Chaihuo, whereas most of the employees were not—marking a class-and-expertise divide that was clearly expressed through linguistic barriers (the inability to speak English) and social distances (the impossibility to create and foster cosmopolitan ties). Affixed to one of the walls was Seeed's timeline with pictures from its early days as an internet shop operated from an apartment bedroom in 2008 to the present company that employed more than one hundred Chinese graduates. The photos reminded me of the mythical narratives of Silicon Valley that were said—erasing early experiences of hardware and software communing in computer clubs—to have their origin in a couple of garages in Palo Alto. The second floor of Seeed's industrial building was occupied by the Agile Manufacturing Center, where small groups of designers and engineers worked on new designs and tested existing ones. Factory workers ran a large pick-and-place machine and inspected boards, performed small fixes, soldered, and boxed small batches of components. Seeed's HR described the company as an "open factory," one that could be "used" by designers abroad to run their projects if they proved to be popular online. On the factory floor, workers could be found performing a variety of tasks from the electronics design of circuit boards to assembly, quality control, and packaging in a well-lit, ventilated open floor space.

A month after my initial visit, I returned to Seeed headquarters with a group of Euro-American visitors on the annual "hacker tour." The same openness I experienced during my first visit was extended to these foreign visitors—to the point that the factory-floor visit seemed scripted to me. Seeed was very well-connected with companies and independent open technology developers abroad at that point. Connections were activated, fostered, and multiplied as Seeed engineers, technicians, and marketing directors traveled abroad to attend meetings and conferences throughout the year. Seeed had been a keen participant in the global open technology circuits but with a specific role and mission as a "convener" for technologists with the goal of changing the image of China abroad as a "copy-cat" by creating local conditions for technological development.

Based on his experiences with Seeed, Mitch Altman describes its observation of social ties: "When [Seeed] wanted to make TV-B-Gone kits [at scale], they asked me first. They did not have to. There is no agreement about it, but it is a nice thing to do. So, we had to ask [Adafruit, an Open Hardware company] too, because they are the ones making TV-B-Gones and because I [had] an agreement with them. I said let's bring them in. We discussed what would make everybody happy. And if they want to call them TV-B-Gone kits, they can give me some amount, a quarter or whatever, and sell it. Instead of making them cheaper but exactly the same and then selling them and competing against [Adafruit] in the US, they would make the Chinese version for all models of TVs in China, including the China-only brands. And it worked. Everybody was happy with that." This arrangement is evidence of that which exceeds open licensing with respect to its (tacit) moral obligation—that is, what is not specified in the licensing terms per se but operates as a moral injunction creating and sustaining ties in common circuits. This moral orientation creates and sustains ties among technologists within the imagined community of practitioners. A gift is inalienable in the sense that it travels with the identification of its initial contributor to the commons of open projects, something that has been described in a very distant cultural context of gift giving in Melanesia as the paradox of "keeping-while-giving" by Annette Weiner.[14] This inalienability of Open Hardware gifts provides the basis for the moral valuations that inform the conversion of gifts and commodities. It also marks provenance for the gift giver as the originator or biggest giver for a particular project or set of community projects. Interestingly, though, this circuitry is grounded in the practice of gift giving among strangers: a condition of possibility for transnational collaboration whose "outside" is comprised of practices deemed "wrong"—that is, not in the "spirit of Open Source."

Derivative projects that do not reciprocate or respect project names and logos as trademarks and that flood markets with "low-cost clones" to compete with "originals" (thus bringing forth the collective orientation toward improvement of the original as an ideal condition for partial but innovative reproduction) constitute personae non gratae. Historically, commercial agreements created in parallel with open licensing of common digital goods (in so-called dual licensing arrangements) have been received with skepticism or have been strongly rejected in Free Software technology circuits. This orientation has been changing in the past decade with the creation of mechanisms for compensating developers and designers (if the person benefiting from the code is not a developer or designer and will not reciprocate with code or hardware design files). It was at this point of inflection that Open Hardware started to consolidate its own community of practitioners outside the Euro-American context.

During another visit to Seeed's Agile Manufacturing Center, Euro-American hackerspace members gathered around Eric Pan while he described the curious trajectory of a product called the Crazyflie. This tiny remote-controlled quadcopter was designed by a Swedish group as an Open Source project but commissioned to be fabricated by Seeed (which, in turn, contracted out locally every aspect of the project, apart from the microelectronics assembly and quality assurance). As soon as the product started to be distributed online with its design files, various copies rapidly appeared in the Shenzhen market, some of which had very clever improvements. As Eric held one of the new iterations of the Crazyflie (2.0) in his palm, he cheerfully recounted the iterative development process to us. First, Seeed released the open project as a product, then it acquired derivative versions that "copycats" had made in the local market. Then, it proceeded to study the improvements made by the copycats and incorporated them into the new version of the Crazyflie—initiating yet another instance of hardware sharing within larger exchange circuits in the broader network. When I mentioned how many times I had heard people praising Seeed for its fabrication services, countering the perception that Chinese manufacturing was "pirated and low-quality," Eric was quick to add: "We try to do things differently here."

Of Copycats and Bad Clones

Shanzhai, or copycats (as it is called in the Euro-American context), is considered an unavoidable reality of the sociotechnical landscape of contemporary China. The term came to the attention of the Free and Open Source community when an influential hardware hacker, Andrew "Bunnie" Huang, first described it in his blog:

> The contemporary shanzhai are rebellious, individualistic, underground, and self-empowered innovators. They are rebellious in the sense that the shanzhai are celebrated for their copycat products; they are the producers of the notorious knockoffs of the iPhone and so forth. They are individualistic in the sense that they have a visceral dislike for the large companies; many of the shanzhai themselves used to be employees of large companies (both US and Asian) who departed because they were frustrated at the inefficiency of their former employers. They are underground in the sense that once a shanzhai "goes legit" and starts doing business through traditional retail channels, they are no longer considered to be in the fraternity of the shanzhai. They are self-empowered in the sense that they are universally tiny operations, bootstrapped on minimal capital, and they run with the attitude of "if you can do it, then I can as well."[15]

In the Euro-American debate, shanzhai is taken as an example of a "world without IP," echoing long-lasting forms of othering and prejudice regarding the "absence of the rule of law" in the mainland. There is, however, much more to it than what meets the orientalist eye. On this topic, legal scholars have spent considerable printer's ink in arguments over the Confucian roots and the political history of China to explain the "impossibility" of intellectual property protection.[16] What is particularly telling about this sociotechnical and commercial context, however, is the unlikely advent of Open Hardware in the hyper-accelerated engineering hub of Shenzhen—that is, the organization of an Open Hardware supply chain as a space of encounter and circulation between mainland Chinese and foreign technical objects and experts creating new ethical and legal sensibilities. As I discuss through the examination of personal trajectories, what is at stake here is a set of practices of production and distribution that rely, curiously, on moral ties for Open Hardware exchange—the same kind of ties that would otherwise form through interpersonal, familial, and more mainstream entrepreneurial practices. What anthropologist Cori Hayden has aptly called the "constitutive outsides" of the digital commons under the neoliberal drive for "innovation" is particularly useful for understanding the demand for "proper copying" in open technology projects.[17] According to Hayden, there is an unresolved but constitutive tension that shapes the lifeworld of the "digital entrepreneur" involving the importance of fostering an electronic commons while placing emphasis on creativity and innovation. Albeit unwittingly, the figure of the digital entrepreneur is shaped by deep colonial and imperialist histories, as we are reminded by studies of digital innovation in the Global South by Anita Chan, Lily Irani, Héctor Beltrán, and Erin McElroy.[18] A derivative cannot be a mere copy; a derivative cannot derive as freely as it narrated to be. It needs to subscribe to the liberal (and imperial) order of the "proper copy," organized around individual rights, obligations, and return on investment of entrepreneurial self-making.

Based on his experience of brokerage in entrepreneurial circuits, Eric Pan discussed shanzhai in terms of an elementary stage of the learning process:

> It is like learning to write; the first thing you do is to copy from the textbook. One would write the [letter] *A* a hundred times or more. I think in China we are in the copy stage. We imitate and, going further, we will stop imitating. "I can do better than you," so I can start exploring things, I can start making things out of my own thinking. So, China needs more time. Copying when everyone else is copying is going to have an impact on business models. I think it is similar to fashion. In fashion you do not have copyright; you have control of the brand and you have to act fast. [People build upon each other's work a lot] and that does not hinder creativity.

Copying is not problematic because it still animates the circulation of open technical objects. It is by getting closer to the moral circuitry of Open Hardware that "proper copying" becomes a moral injunction. Shan-

zhai operates outside open technology circuits, despite their family resemblances. The "constitutive outside" of Open Hardware circuits—the
practices of cloning, for example, that do not observe moral ties and legal
parameters—is of particular interest to us as it puts into evidence technical, moral, and political sensibilities as prerequisites for meaningful participation in the community. To this we must always add another aspect of
the "constitutive outside," which is composed of all of the "makers" who
are not made visible because they occupy precarious, low-paid positions in
contract manufacturing companies.

Echo and Narcissus in Postsocialist China

My host and I were sitting, as we usually did, in his living room, which had
been converted into an electronics lab and stock room for his online shop
in the periphery of Shenzhen. Piles of blue bins holding a wide variety
of electronic components could be found next to a mid-sized laser cutter
whose ventilation tube came out of the apartment window. Right next to
it was a soldering station with spools of colorful wires. My host was not on
good terms with his neighbor, who demanded that he stop cutting plastic
because of the fumes expelled into the main courtyard of the working-
class apartment complex—exactly the same complaint from nearby shops
that had prevented SZDIY members from using the laser cutter at Chaihuo.
His finished and unfinished prototypes were everywhere in sight: small
robots, electronics kits in plastic bags, extension boards he had designed
himself, cardboard boxes, breadboards with half-assembled circuits.

Pointing to his computer screen, my host showed me a Taobao website, a Chinese portal for webshops in which most of the domestic electronic transactions for small to mid-size quantities gets done. On display,
he had the shop of a local open technology enthusiast showing the logo
of the Open Source Hardware Association (OSHWA) with a picture of an
Arduino clone. My host laughed and said, "People [are] not doing it right"
by just copying an Arduino, which was made open in its licensing terms
with a simple but key restriction. Replicating the hardware (and its documentation) is allowed if the brand is removed and the derivative project's
name does not cause confusion with the trademark Arduino. It took only
a few microseconds for the moral justification to arrive: "But [the owner
of the webshop] is selling to the Chinese [market]. . . . He just had a baby."
My host followed with a story of how good this particular engineer was

but also how far removed from Euro-American debates about IP and open licensing.

In a couple of weeks after this conversation about open technology, I began interviewing Chinese founders, software developers, and engineers who had experience with Free and Open Source projects in Shenzhen. One of my interviewees introduced himself as Wei and invited me to come to his office in the region of Bao'an, not far from where I was staying and where I often witnessed poor city dwellers being harassed by the police when crossing city limits. Shenzhen is a place with highly competitive managerial and well-paid IT positions. Its proximity to Hong Kong also makes it a place where those who have permission to leave the mainland can foster cosmopolitan connections. As one of the cities with a well-guarded Special Economic Zone, one needs government approval to stay, the hukou (household registration), which is heavily enforced for rural migrants.

Wei's office was located in one of the multiple industrial enclaves in the city, where there were numerous workshops for wood, plastic, and metal work, as well as several companies providing all kinds of fabrication services. These enclaves are surrounded by residential apartments, big shopping malls, and small businesses replicating a modern architecture to be found in other parts of the city. Wei figured he had lost his office key, so we waited for the locksmith. I stood there quietly thinking that our hackerspace friends would be more than willing to help open that door for free—myself included, but I also wanted the locksmith to be compensated for his craft. When we eventually got into his office, Wei told me casually that he had started his own business because of the pressure he suffered from his wife's family. He did not mention the baby, but I knew the story from his old SZDIY friends. Wei was one of the Seeed Studio's first engineers and experienced its growth from a few employees up to a hundred in the process of creating ties with foreign companies, engineers, and designers abroad. He thought highly of Seeed as the place where he came to understand Open Hardware: "When I got to play with Arduinos, I realized it was completely different; it was so convenient. The most amazing thing is that an Arduino is built by people on the internet and it is meant for artists. The most important thing is Open Source. Seeed used the concept of Open Source so well. They created Seeeduino [an Arduino clone] by collecting ideas. They asked the community: 'Is there any improvement we can make to the Arduino board?' So, Seeeduino was based on ideas from the community. It was accepted by the community. Seeeduino was the first to become famous. Because of [their] example, I came to learn about Open Hardware. Its power is very strong. It is different from the traditional way of designing things." Seeeduino contrasted with the so-called bad clones that we could find in China and everywhere; it fulfilled the condition of improvement of an existing open project, returning the contribution to the commons in order to be considered a legitimate derivative.

Small companies in China such as Wei's online shop had to move as fast as possible in the cutthroat environment of Shenzhen, and this sense of urgency was one of the main reasons for the widespread practices of fast hardware cloning. Interestingly enough, economies of technical services and gift giving, such as the ones involving open technologies, represented to Chinese engineers a transformative experience through an ethical reflection on copying practices. Under what circumstances to copy? How to copy? What was meant to be copied and what was not? The experience of entering the global exchange of open technology meant for Wei and other Chinese entrepreneurs of this generation, such as Eric Pan, slowly coming to situate themselves in new circuits of participation, learning to learn about the technology and the moral sensibilities concerning copying and cloning, figuring out what they could improve and what they could return as a gift. Under what circumstances should they ask permission to run a project on industrial scale for profit on their website? Learning how to interact and position themselves in the context of open technology production went a long way, as expressed by Wei about his experience as a Seeed employee: "For me, what I can do is try to make things open and follow the licenses at least in the international market. But here in China it is difficult, to be honest. I keep hope alive for change here. As the economy develops, the idea of licensing will change. In terms of personal experience, people are always trying to make something acceptable and respected by the community, not just a copy—even copying without attributing the original. . . . Things in China are a little bit different. You know, some people are trying to make money as fast as possible, and the fastest way is copying. I do not think it is sustainable over time, since things are changing. China is more and more open. But there is a long way to go. Why am I so hopeful? People always want to be respected. Once they get better, they will try to make something valuable, as valuable as the new things designed in America."

Copying without regard for licensing restrictions and moral injunctions for attribution and sharing was the norm as mainland companies were more secluded from watchful and litigious foreign eyes. The context of production, distribution, and usage was perceived to be changing, however, and the change came about with force as Seeed and other companies started to create strong ties with companies and engineers abroad. During our meeting in his office, Wei showed me his website where the Arduino clones he made had the Arduino logo blurred. He then told me he was complying with a new regulation by the online infrastructure owner, the internet giant Alibaba, to enforce trademark laws by forbidding webshops to display and sell products that infringed trademarks. A couple of months after we first met, I returned to his website, which showed similar Arduino clones, extension boards, and accessories, but now with his own brand name and logo silk-screened on his products.

Wei was inspired early on by the magical arguments put forward by Chris Anderson in his popular book *Makers*, depicting Open Hardware initiatives involving fast prototyping, 3-D printing, and successful open platforms leading the world to a "new industrial revolution."[19] This was the same Anderson that would make uninformed claims about the "end of theory" with the rise of large-scale data analytics. Wei was hopeful about seeing the realization of Anderson's entrepreneurial vision. He wanted to take part in the promised "revolution" with other Open Source companies, such as Seeed: "It is a good thing to be open, be honest, and make things—to make hardware as well as software projects. The market is getting bigger and bigger. Customized products are becoming more and more important. This is why there are so many small companies to meet small requirements. I got to know from Chris Anderson that the third industrial revolution is really happening here with Open Source Hardware, 3-D printers, Arduinos, and the community."

Entering cosmopolitan relationships through open technology is not trivial because it involves a different orientation regarding ownership, property, and "propriety" of one's technical and commercial conduct. "Openness" relies upon moral circuitries with specific associated milieus. In China, economic hardship is often ranked as the prime motivator for the use of Open Hardware; it figures often as a means for short-term economic return. This crucial socioeconomic distance challenged Euro-American dispositions regarding the "right way of doing Open Source" by emphasizing the conversion of gifts into commodities. To seek economic benefit and competitive advantage to the detriment of the spirit of Open Source was hardly unique to hardware entrepreneurs in Shenzhen, however.

I have been told in private by Open Hardware engineers about disloyal practices of unfair price monitoring that never surfaced in public debates but are known among Euro-American community members. Conflict over the blurred boundaries between gift and commodity forms serves as an important index of the moral circuitry at work. One of the most active participants in open technology development remarked that "there have been times when people were not so happy with [a company] copying their projects and selling in their website. [I have heard about someone scraping a competitor's website to monitor prices] and getting busted and then doing it again. It is just violating the spirit of the whole thing because the spirit of Open Source is sharing, and if we all do that, there will be enough for [all of us] to make a living doing what we are doing. We do not have to make zillions of dollars."

The passage from cloning trademarked Open Source Hardware to blurring or masking trademarked logos on public display, to finally adopting one's own logo for redistribution and commercialization of improvements to an original design constitutes a large-scale transformation toward what anthropologist Lisa Rofel identified as "cosmopolitanism with Chinese characteristics," after China joined the World Trade Organization (WTO) in 2001.[20] In the past two decades, a major shift in economic policy toward market liberalization has brought the People's Republic of China (PRC) closer to the United States as a major trading partner—an ambiguous position, Rofel reminds us, of competitor, collaborator, and protagonist of an ongoing "technology war." Political economy debates concerning China as a new imperialist force in Southeast Asia, Latin America, and Africa attest to this recent and unexpected transformation with major geopolitical consequences. Chinese patent applications skyrocketed in the past decade. Major internet companies, such as Alibaba and Tencent, were pressured to comply with the WTO Trade-Related Aspects of Intellectual Property Rights (TRIPS) agreement and create mechanisms to enforce intellectual property. Interestingly, most infringements were related to trademarks, followed by copyright due to distribution of films and software. The situation with patents, design patents, and trade secrets was even more nebulous, given the challenges of traceability. Where can we start to determine the source of an invention? How can it be assigned to an individual (or group of individuals) when the "novel technique" or "original expression" to be protected draws from the work of many others?

From a Euro-American standpoint, Chinese copyright law was considered insufficient for the maximalist purposes of the WTO TRIPS agreement, so amendments had to implemented. One document to this effect was the Regulations on Computers Software Protection from 2001, which contains provisions for software copyright in China. By 2013, the Creative Commons suite of flexible copyright licenses (in its version 3.0) was fully translated and adopted in the PRC jurisdiction, rendering many open licensed projects legally enforceable. As of the date of publication of this book, no official Mandarin translation of the GNU General Public License and other Free Software licenses were in circulation, although local activists at SZDIY were privately discussing translating them. Other licenses denouncing inhumane labor conditions, however, have become much more prominent in the meantime, such as the "Anti-966" license in the emergent category of "ethical licensing." They fundamentally occupied the space for ethical problematization that open licenses were not originally intended to cover: the use of open technologies in human rights abuses, the exploitation of volunteer labor, corporate accountability, and anti-racist action.

In the intimacy of Euro-American technopolitical projects, China used to be Echo, but it is now becoming Narcissus—to borrow the characters of the Ovidian epic *Metamorphosis*. In a geopolitical IT war, the United States entrepreneurial class has had quite a bit of difficulty finding beauty in an image that is not its own, whereas China has found itself relegated to repeating what Narcissus says and does. Core-periphery relations started to look different, in time, with a new international division of labor around design, manufacturing, and distribution of everything that infrastructurally supports the hackerspace network. Narcissus considered himself too precious to get involved in low-paid, "little-value-aggregated" labor-intensive tasks, which he outsourced to others in the domain of manufacturing, packaging, testing, fixing, and shipping. Echo was too enamored with Narcissus to question her role as the mass replicator of the most enchanting and yet superfluous and ecologically disastrous electronic designs. Eventually, the relationship turned sour: infatuation is a wildfire that does not burn for very long. Echo decided to take up the position of designer herself: China started to invest massively in its own high technology sector with hopes of creating a prosperous "knowledge economy." In his reading of the Ovidian myth, Jacques Derrida tells us that Echo eventually makes Narcissus's language her own in a spectacular reversal (making the curse of repetition into a possibility of new expression) that is

motivated by love[21]—the love of technology through which Chinese technologists and designers increasingly make Euro-American projects their
own. We can see a parallel in how Chaihuo echoed hackerspaces abroad
until it started to develop its projects through its cosmopolitan circuits.
With a convivial and independent spirit, we heard from SZDIY members
as they echoed other hacker projects until they started to develop their
own voice. Copying (as repeating but within a new circuitry) to be able to
say something new and echoing (as identifying and mirroring) were the
only possible ways of forcing Narcissus to attend to someone who was not
himself (or a copy of himself). It was about time for the technical landscape to change, setting a new stage for alterglobalized circuits. In the infinite regress of mirror images of copies of derivative copies in the Global
South, open projects materialized new technopolitical possibilities, but
not without inequities that still abound in common circuits.

FOUR

NalaGinrut

Hacking as Spiritual Calling

NalaGinrut, or Nala, whose nickname is Alan Turing spelled backward with the repositioning of just one vowel, was a key figure in the Shenzhen Do-It-Yourself (SZDIY) collective. Known for his very strong programming skills and organizational capacity, he was organic to the open technology community in Shenzhen, as well as a keen activist in the Free Software movement within and beyond China.

Nala was introduced to computing early on by his paternal uncle, a computer technician at the time of the debut of personal computers in the Chinese market in the early 1990s. NalaGinrut was eight years old when he started playing computer games with friends and sharing software. Concepts of copyright and improper copying were completely absent from the horizon of his early digital experiences. What really mattered for him was the ability to run games and share the experience with others.

Nala grew up in an urban family in the city of Kunming, Yunnan province. His parents were highly supportive and caring, pushing as hard as they could for his education. He recalls being a serious student in elementary and high school because of familial pressure. His mother was born in Kunming. His father came from a small village in Yunnan and worked for a time in a copper mine before he found a position in the army that allowed him to move to the city. It was there he met Nala's mother, who held a job at a bank at the time. *My parents were traditional and affiliated with the government. I got very important experience from my father. He was born in the countryside, in a very poor place, but he managed to work and leave to get a better job. He had no education. He taught himself pretty much everything, reading articles and organizing people's resources, doing management; he climbed up step by step for me. Unfortunately, he worked for the government. At the time, it was his best option. I can say I respect my father because lots of people respect him; he worked for the government doing clerical work, which made our lives easier.*

Nala studied management science as an undergraduate at Chongqing University and had the experience of living in a dormitory with students from rural parts of China. During his undergraduate years, he came to realize the financial struggle of his colleagues, who depended on scholarships to support themselves. Nala was not studying to become a computer scientist or IT professional at the time; he occupied himself with other hobbies, which included playing soccer and classical music. *I was aimless during my university studies, I did not even know what I was learning at that time, although I got an "A plus" in the final evaluation.*

After graduation, he decided to take a very competitive postgraduate-school entrance exam, so he rented a room with the help of his parents and studied math and computer programming in complete isolation. His immersive experience with programming started fairly late in comparison to other senior programmers, such as Mitch Altman or Gniibe Yutaka. He applied to graduate school in automation engineering, a degree combining computer science and electronic engineering, and he concentrated on the study of C and later Scheme. *Before learning about Free Software, I was an enthusiast of computer science. My curiosity and my mind couldn't stop from learning about computers. The chance to learn with Free Software revealed that this way is the right way to achieve technical progress. It is a love I can't stop.*

During his postgraduate years at University of Electronic Science and Technology in Chengdu, he participated in the Linux User Group (LUG), where he learned about key Free Software projects, such as Linux, Emacs, and Guile. He recollects conversations with his professors about Free and Open Source technologies, but they had no idea what they were. *My professors always encourage me to dig deeper into the technical topics and do my best to find my position in the industry. They never have heard of Free Software. I remember clearly. It was the time that Stallman first appeared in Chinese media. It was too early for my professors.*

His focus on self-directed learning and participation in online communities with experienced Euro-American hackers gave him a distance and perspective from emphasis on examination, certification, and status-seeking in his educational experiences.[1] From his critical stance, Nala approached the question of hacking by drawing a distinction between engineers and hackers, or in his parlance, between those who are limited in their knowledge by virtue of their academic training and established industry practices and those who have a wide range of interests and accumulated knowledge in computing, being therefore able to engage in projects of higher technical complexity. From his experiences interacting online, he came to identify people in open technology projects as sources of inspiration. He spoke highly of the mythical MIT AI group of Lisp hackers and said with joy that *Structure and Interpretation of Computer Programs* by computer scientists Gerald Jay Sussman, Julie Sussman, and Hal Abelson[2] was his "bible," which he studied thoroughly to teach himself programming. He admired Lisp programmers and characterized them as "real hackers." After spending time alone teaching himself to code, he proceeded to study Scheme, dedicating time on mailing lists and on the IRC channel for Guile. *In the Free Software community, what I learned first was the skills. Of course, that is what I was urged to learn from the real hackers. And I have learned how to talk and cooperate with hackers from a technical standpoint.*

One of his personal goals was to help foster a hacker community in Shenzhen. Based on his transnational experiences with Free Software and his aspiration to foster a "hacker spirit" among local engineers and software developers, he promoted hackathons and "hardware freedom" events, gave keynote presentations for the local Software Freedom Day, and more recently, organized the creation of a local community space, the SZDIY hackspace. *My friends and I are guiding the SZDIY community to the correct direction in this dark environment* [of Shenzhen] *and trying to keep our hacker spirit alive. I don't play with Open Hardware just because it is hot in the market. I want to let hackers have the freedom to hack and share, so they can survive in this world with their skills. That is what I am fighting for.*

Nala's perception of formal education has led him to search for alternative ways to teach himself computer science. During his graduate school years, he witnessed a memorable if not traumatic event in which a sixteen-year-old programmer announced to the technical community of Nala's school that he knew several programming languages and could address any technical question. *The result was a flame-throwing fest. High diploma guys bashed the young student, directing bad language at him because of his age and his claims. At this age, you have no authority to say such things or claim to know or have learned so many things because many on the other side are researchers and high-ranking people. They can do it as well. The guy was beaten! That was the feeling I got at the time. If someone so young is treated like this, I thought, maybe I should go away, I should not stay in such an environment anymore. This is why I liked the Guile programming language community. I was a naive learner at that time, but the community members were very friendly and taught me many things. I felt this was the community where I should stay. For beginners who know very little, it's important to find a place where people treat you well and want to teach you things. In the [local] technical community, there was no such attitude.*

Before reorienting himself toward a different career path in computing, one that was carried out through commoning practices, he took the route of formal education, coming to participate online later in practices of self-cultivation through online debates. *My thought is simple: if you want to be a hacker, you have to be around other hackers. And someday you find you can contribute naturally.* He would spend countless hours on IRC and read books and papers, search, and get pointers for technical information, being encouraged by some of the most active members of the Guile community.

His organic involvement took roughly three years to put him in a position to contribute code, and he was praised for the contributions he gave over time, which provided even more motivation to persist and pursue a job as a developer working remotely for a German Free and Open Source company established in Beijing. He continued to be actively involved in the regular meetings of the SZDIY and the Shenzhen Linux Users Group (SZLUG), having helped to bootstrap various events and a shared laboratory for the community. *It is amazing to have such communities in Shenzhen. Everyone here is a reasonable person. Nobody is an idealist. This city is very practical; you have to do real work for money.*

The practice of giving software gifts (lǐwù) assumes different forms for the Free Software community in China. *If you help me with a trivial thing but it is really important to make me overcome a certain difficulty, Chinese people would treat this very seriously. We use an old saying to express this: "One drop of grace must return a thousand drops"* (滴水之恩当涌泉相报). *The situation* with Free Software is different because most Chinese technologists don't think they have the ability to contribute back, even though many of them are willing to advocate Free Software and help others try Linux. We've been comfortable with this for the last five years. When we contribute what we can, everybody can get involved. I think we are experiencing an "Age of Enlightenment" for the Chinese Free Software movement.

Before landing a well-paid position as a software engineer in the research and development division of a German company, Nala held technical positions in small and large private and public companies. Fresh out of graduate school, he worked as an intern for ZTE in the sector that was dedicated to mobile communications. His interest and knowledge of Free Software was so well-known among his coworkers that he would often be visited by engineers from other sectors to troubleshoot GNU/Linux systems. Working under a boss he described as "clueless" and with technologies that had "no future" led him to give up his job. His boss often dismissed his technical interests as a "pastime for hobbyists, not suitable for the industry, a waste of time." Nala used money he had saved to support himself while he worked on his own project, a small operating system written in Scheme. Luckily, he could count on the strong support of his partner at the time, who later became his fiancée, despite, he says, his "uncertain" professional prospects.

Nala attributes this experiment with independence to his willingness to make sacrifices in order to keep his hacking time alive. *My hacker life forces me to avoid letting it die; I would do anything for its survival. You know, another hacker once told me that he was born a hacker, so he never considered losing his hacker life. I asked him, "But what if someday you lose your hacker life?" His answer was interesting: "My hacker life forces me not to lose it, and I never will, and I never will. . . . I would do many unbelievable things for it." The spirit is very important for a hacker because sometimes you must have a very strong reason for the things you do. Sometimes you do things 90 percent and some weakness will occur in your mind, so you must have a very strong spirit to face such weakness. It is a regular practice, an everyday belief.* Importantly, the conception of a "hacker life" takes precedence over other domains of social life as well: *Sometimes my hacker life pushes me hard. I am sleepless. You know, the guys in the Guile community have the same problem. Sometimes I see people on the IRC and I ask, "Is it very hard for you to sleep?" And they respond, "I can't sleep, I have to hack on something." All of us have the same problem about sleep. The extreme case is* [one of the key contributors to the Guile project]; *he told me his sched-ule was highly optimized for his hacking, which is far from normal. He hacked himself, I think.*[3]

Nala's second job, a temporary contract with a small embedded-systems manufacturer, was not exciting or challenging for him, but it was considerably better than the one he had at ZTE because he got to work on his own GNU/Linux operating system, which he implemented "from scratch": *I started to get involved with Guile, write Guile code, and started to work on my operating system project as a night job. It was the only hope, the only happy time I would get. My partner and I had less money, but it was a happy life when I got home. I do not think I got so many things from company I worked for because, you know, I do not think that a company doing embedded hardware gets to improve every day. If they can sell things and get money every day, OK, that is enough for them. There was no improvement. I learned industry management as an undergraduate, so I knew their production process was nice, strictly controlled. It was good, but for an R&D guy, it was not so cool. Many companies in Shenzhen are in the same situation. There is a good manufacturing process, but no technical skill. They have good products but less technique. So I told my wife that by the end of the contract, I would get 24,000 yuan, enough for my operating system hack for a month—I mean, for several months. To which my wife said, "So do that!"*

When Nala's contract expired, he came to talk to his boss. The company had no long-term plans for him, so he decided to leave. Throughout this time, he kept studying the topic of operating system design inde-pendently. After quitting his job, he gathered all the information from the OpenCourseWare at MIT on operating systems (6.828 "Operating System Engineering") and dove into his project, not as a side project as it had been before but as a full-time one while he was unemployed. *I think I have a real knack for teaching myself. Well, maybe that is one of the reasons why I am not a regular engineer.* After two months of unemployment, he found out about a job and submitted an application to a Free Software company, which ac-cepted him without much delay. Some of the employees at the company knew about his work. He hypothesized they probably learned about him by performing a web search and finding his public repository with a copy of his unfinished operating system.

Nala dedicated himself to the task of informing the Chinese IT com-munity of the importance of looking elsewhere as well for the history of their technical interests. "Maker Culture and Anti-Consumerism,"[4] a text he published with a friend, a teacher in one of the local elite schools of

Futian, Shenzhen, and core member of the SZDIY group, offered an interpretation of the political potential of the maker movement that was quite
distinctive. Their narrative drew curiously from the genealogy of hacker
collectives in Europe, such ASCII, monochrom, and Dyne, given their
commitment to the anti-corporate globalization movement, but also from
the anti-consumerist potential of the Open Hardware community (Danfei
and Nalaginrut 2014).[5] According to the authors, these movements culmi-
nated in the new figure of the maker (chuàng kè) through the combination
of DIY and anti-consumerist practices to repurpose and reinvent existing
technologies.

Nala's activism for Free Software is expressed in his search for ways
to foster a local community with the help of other members of the SZDIY
group. In his vision, it is necessary to align commercial interests of companies with the activities of the local community. *Many good CTOs, founders of companies in China, may get the meaning of Free Software. I always try to
show them a different way to contribute to the community. If you tell a business
person, "OK, you use Free Software, but you have to pay back," they definitely will
not hear you. But if we choose another way: "OK, you try this Free Software and
you get profit. OK, your products are well priced and you get good money; maybe
donate some money to the community and you get prestige." Why would we approach them like that? Because their education is like that. If you want them to
make some technical contribution, well, they just can't. But for money, they can
buy reputation. Today many rich Chinese men pay for reputation.*

The software and hardware development strategy with open technologies is also meant to address the issue of copycats (shānzhài) in China.
*This kind of culture is very bad. If you expect people to do creative things, that is
unrealistic. Very good engineers go abroad to be regular company engineers. The
environment here is very bad. No joke. Maybe people are just trying to survive.
The irony is that while these really outstanding engineers go abroad, others who
are not outstanding are rejected by universities or research institutes here. The
[local] IT companies will just copy. This is a very important thing: we do not have
a community to lead people to explore their inspiration and hacker spirit. The only
way I can contribute to solving this problem is to try to build a good community.
That way we will let outstanding individuals develop their aspirations. If we do
not have outstanding people, our spirit will die, no doubt.*

Fighting the widespread perception of hacking as criminal activity is a
challenging local issue as well. The use of metaphors over the internet is a
common technique to avoid filtering and censorship. *In China, many words*

are polluted, like "young lady" or "mistress" (xiǎojiě), which now means "pros-
titute." Another word is "hacker" (hēikè), which is specifically used for crackers
(hàikè), illegal crackers. Universities will teach students again and again, "Do
not be a hacker! It is very bad!" Free Software and Open Source, especially Free
Software, in China have such a hard time for this reason. People think, "Well,
freedom, whatever you want to do, is OK." Freedom in the Chinese context may
mean that you do not have to pay it back or contribute. Just take it. So, you know,
many people do not know they have the duty to contribute and pay back. Not to
mention that software is not free. Free Software is not free. People can pay for it.

The question of retribution is much more difficult when it comes to
gift giving among strangers.[6] For Nala, "hacking" involves an ethical
demand that extrapolates the practice of software development or hardware design. *This is a metaphor not only in software but for everything in my life.*
I'm seeking it everywhere I go. So what is real freedom? It's not doing anything
you want. That's not freedom; that's self-indulgence. Real freedom is doing (or at
least being able to do) things you should do. If I want to share knowledge with my
neighbor, this is what I should do. Jesus told me to love my neighbors. If some laws
or rules prevent me from doing that, they break my freedom. It is just protecting
the profit of someone else, although duped people always think this is right.

For the members of his generation growing up in the context of Deng
Xiaoping's economic reforms, politics became an unbearable topic: *I think*
I have no special interest in politics. Many young people in China have heard too
much about it. They just hate the word . . . too much politics. When asked about
his work for Free Software and its political relevance, he would recon-
sider his position and cast it under a different light: *When I get involved in*
Free Software, maybe I am just playing the role of a political figure, an activist.
Well . . . you know, I can only answer this question in a religious way.

Nala is a devout Christian and identifies as a New Calvinist. His church,
however, is not a church: it looks more like a person's house. It is a big
rental room near Huaqiangbei where they assemble for Bible readings. His
priest knows the official state religion representative well, so he figured
out a way to create space for their religious practices. When he was growing up, there was no discussion about religion in his household; it was
identified as a "source of evil." Even to this day, the topic of his newfound
religious belief is not discussed in his family. It quickly became a taboo
after an initial conversation he had with his parents, after he was already
a convert. *I have a duty not to just mention Free Software to the other people. I*
have the duty to practice, to lead people in this land to learn about it. My duty

is to be the messenger. I am not the hero because there are too many people in Free Software and Open Source who are heroes. They can influence many things beyond our expectations. I just cannot do that. Maybe my kids will be able to do something more than me. For now, I am just a messenger. I have to work, to practice, and do something for other people.

When questioned about his motivations for engaging in hacking, Nala would elaborate on the importance of creating a new future for his children. *What I wanted to tell you is . . . why am I hacking! All the knowledge, all the books I have read are meant for hacking for the children. Yes, for me, the final hacker purpose is the educational purpose. I cannot say it is for all educational purposes or for other kids, but it is for my kids. Someone once told me this maxim: "The most valuable thing a father can give to his kids is his experience about life." Not assessment or money or a house—they mean nothing.* Nala's overall views on hacking are often explained with metaphors from martial art novels. One subgenre in particular describes the virtuous martial art of ancient China, wushu (武術): *The metaphor of wǔ means "to stop fighting." It describes the tools and functions of fighting: fight to stop fighting. This one here is a very important word for Chinese people: xiá. Not everyone who knows martial arts can be called a hero, but those who have a good reputation and very high kung fu can do something to stop bad things. So you can imagine Stallman: he could have been very rich because he had very good skills; he could've hired the highest kung fu masters, but he was not in it because of an ambition, to earn something for himself. The wǔ metaphor evokes the man who strives while not standing out. He is dedicated not to be outstanding, despite the fact that he is really outstanding. The metaphor contains a very profound philosophy of what it means to be a real person, because if you are really outstanding and you try to show how outstanding you are, you will have few true friends. The art of xiá is the antidote to this. You will get many friends and you will not have big pressure to find work or pressure on your hacking. This goes hand in hand with wushu. These are the metaphors I use.*

Nala has continued to work on his open technology projects, but he is no longer connected with the activities of SZDIY on the ground. He relocated with this family, continuing, as he puts it, his "journey" elsewhere. On his migratory route, he carried two books with him: one that he had already mastered, the so-called tiger book, a highly technical volume he used to implement a Scheme compiler for the STM32 microcontroller platform, and another that is still an object of study: Hans Jonas's *The Imperative of Responsibility: In Search of Ethics for the Technological Age.*[7]

FIVE

Tokyo Hacker Space

Bricoleurs Respond to the Disaster

On March 11, 2011, a magnitude 9.0 earthquake shook the northeast coast of Honshu, Japan, and was followed by a devastating tsunami, causing more than twenty thousand casualties and washing away coastal villages in the prefectures of Fukushima, Miyagi, and Iwate.[1] Televised damage and ineffable human suffering was not limited, however, to the coastal areas. The magnitude of the earthquake shut down the nearby nuclear power plant, Fukushima Daiichi, which was soon after struck by a twelve-meter tall tsunami, knocking out its power supply, and severely impairing its nuclear reactors. Explosions followed as three out of six reactor cores melted down due to a cooling-system failure. As an urgent measure, Tokyo Electric Power Company plant managers authorized the opening of containment vessels, releasing thick radioactive steam inside reactor buildings. As if the atmospheric dispersal of radioactive plumes was not already tragic enough, an estimated 1.3 million cubic meters of contaminated water flowed into the ocean.[2] The Fukushima nuclear disaster escalated in a matter of days from a "severe accident," as defined by the International Nuclear and Radiological Event Scale, to a "major accident," the biggest since Chernobyl in 1986. From a historical standpoint, the nuclear crisis spoke loud and clear about the limits of technoscientific knowledge of late industrial societies.[3] It showed us the inverted image of a mad dream of total control: the ability to imagine and dramatize but not prevent the onset of dystopian futures. The nuclear tragedy brought to the fore yet another socio-environmental crisis, and we found ourselves lost in the thick of it, experts and nonexperts alike.

Four months after the onset of the "triple disaster," I returned Japan to continue the work I had begun with preliminary field research in 2010. Searching for meet-ups on the website hackerspaces.org, I found Tokyo Hacker Space (THS), a hub for technical exchange among local and itinerant artists, engineers, and computer hobbyists. By attending their regular meetings, I started to learn far more about radiation than I expected from a group of digital aficionados.

In parallel with other hackerspaces, such as Noisebridge and Chaihuo, presentation of self at THS was an important entry point to its interactional modus vivendi.[4] During "open nights," introducing oneself served to signal collaborative potential, as well as providing minimum regulation to the open-door policy. Newcomers usually presented themselves through a constellation of technical objects to establish commonalities with other computer enthusiasts. Just as they did in other community spaces, co-participants would talk about low-power wireless networks, 3-D printing projects, Arduino clones and extension boards, "homebrew" arcade projects, Free and Open Source software programming frameworks, systems' "penetration testing," and several other information security practices.

I first approached THS with a friend from a Japanese Free Software company specializing in networking monitoring services. We found THS by accident after reading online that every Tuesday night, following the "hackerspace design patterns," the space had its front door open to the public. My friend and I decided to go one night after a bit of hesitation. We arrived slightly late because we miscalculated the time it would take us to walk from the metro station to the space in the upscale neighborhood of Shironakedai. When we arrived, it felt as if we were at the wrong place. THS was in a house like any other in the residential area, except for what was taking place inside. We had to squeeze through a half-open door to avoid calling too much attention to ourselves. The main meeting room was crowded. It accommodated semi-comfortably about twenty-five people, among whom there was no shortage of technical expertise. Following the protocol of an informal round of self-introductions, newcom-

ers took turns highlighting the relevance of their projects and establishing
some level of connection with others, managing impressions as they proceeded. English was the lingua franca, but Japanese conversations and
isolated expressions would be used often in code-switching. More prestigious and seasoned members could be readily identified by the frequency
with which they joked about technical topics, based on what other members were presenting.⁵ The atmosphere was mostly light and convivial. For
every technical or artistic project discussed, a certain degree of openness
was made evident, signaling a potential for exchange.

When it was our turn to introduce ourselves, my friend started, in a
shy, trembling tone and an economy of sentences, to tell the attentive audience about his work on network monitoring. This mode of technical-
object-oriented sociality among strangers, an institution for hackerspaces,
was somewhat torturous for him. It was not only because he had to use
English in public, which was a challenge in itself, but there was something
about having to present his competencies in public that was deeply em-
barrassing. I was not particularly happy to see him put on the spot; after
all, we had not come for this kind of trial. I waited for my turn and then
launched into a quick story about my experiences as a computer technician, my Free Software volunteering work, and my detour into the human
sciences. I was a bit nervous as well, since English was not my language,
but I was at the same time proud of the common experiences that brought
me to the space as a reformed technologist qua anthropologist. I explained
what I came to do as an ethnographer, introducing a good degree of en-
tropy in the mental space of my interlocutors. "Anthropology isn't science
or engineering, right? So why are you interested in Open Source?"

I would often get this boundary-guarding question and would always
answer in two parts: one matter-of-fact and the other with a bit of technical humor. This was, after all, an important opening to understand how
people perceived me (based on how they perceived themselves, their competencies, and their work). I would explain first that I was interested in
understanding how people learn the skills they need to participate in collaborative projects. But I also often said that I was trying to discover an
antidote to "flamewars"—that is, to find a way to help heal online spaces
where people just "flamed" each other for having disagreements on technical, social, and political matters. I was quite naive when I started my
research—and that served me well, methodologically speaking.

Tokyo Hacker Space was conceived at a BarCamp[6] event in 2009, when a group of expatriates decided to create a hub for bringing artists, engineers, and IT professionals together. In his informal role of "hackerspace ambassador," Mitch Altman happened to be visiting Tokyo and registered to give a talk about one of his favorite topics: how to start a hackerspace. Mitch worked his magic with his usual charisma, and many participants in attendance fell for the enchanted prospect of creating their own space. It took only a few minutes before they had a domain name, tokyohackerspace.com, which they registered right on the spot. The easiest part was done. Now it was a matter of finding other people to join and endure the challenges of community building in the megacity.

From the initial group of BarCamp participants came the idea of renting a more permanent space. At the time, prospective members had all been in circulation, attending computer and robotics gatherings, but a more perennial space was felt to be needed. Tokyo was extremely challenging in terms of rental space, especially for an independent group of computer *otaku*.[7] They decided to publish an announcement about their intention of starting a hackerspace, an experiment to see if people would come by, and after a couple of months of holding meetings at restaurants, they had enough supporters to move ahead with an expensive rental contract. In the summer of 2009, the first physical location for THS was arranged in Shirokanedai. It could not have been a better start.

Just like other community spaces in global cities, THS had relatively few square meters to host meetings and projects. Given the way the activities were spatialized, it was impossible to be there and not be noticed. I came to understand over time, in contrast, how often people felt invisible when differences in technical expertise as well as in gender, ethnicity, and nationality were apparent in one's body and language. In contrast to Noisebridge, keyed access to THS was regulated. Technologists who paid membership fees had controlled access, whereas nonmembers could visit only during public events, such as workshops or open house days. Similar to Chaihuo, there was no formal procedure for accepting or declining membership applications. Decisions were made ad hoc and carried out by a small affinity group composed of founding members who were active in the space and upheld the legal responsibilities for the rental property. Among founders and admins, there were only expatriates from Euro-American countries with technical backgrounds, day jobs, and, often, Japanese families.

A membership request had never been declined in the history of the space, but there were problems with newcomers who were perceived to be "disruptive," as always happens in hackerspaces everywhere. One member in particular was the reason the admin group decided to convene closed meetings to decide how to handle the disruption. The "undesirable" participant, a North American expat, would religiously come to the meetings and argue with other members over political issues from a self-righteous militaristic perspective. He was visibly lost in the space and without any sociotechnical bearings. It was as if he had no comprehension of where he was. He would ask other members to fix things for him, an elementary but serious breach of protocol in the space where one was supposed to dedicate oneself to self-directed learning with help from others. As far as the tacit contract among hackerspace members goes, one was supposed to at least try to fix a technical object, even at the cost of breaking it. Members would frequently stop working on their projects to help others, regardless of the problem, generating a virtuous circle of gift contributions with an outward-expanding radius. Being overly dependent on other members to perform technical tasks, however, was a serious breach of the tacit orientation of the group. The shared assumption was that the conditions for learning to learn were given both materially and symbolically around the space and online. It did not take long before the newcomer was asked to leave the space for good.

The spatialization of the activities in the house in Shirokanedai followed key hackerspace patterns. As we passed through the main door, we left our shoes in a disorderly pile by the threshold before coming into a big living room with couches, chairs, and plenty of table surface for displaying projects. The tables were placed against each wall, except for one where video projections were made. In an adjacent part of the living room, a long electronics bench with soldering stations, tools, and parts could be found, as well as a large metal archive cabinet where electronics and computer parts, free to hack, were stored. The cabinet was decorated with stickers from hacker projects and conferences, shared symbols for the community. A lot of the electronics work took place through informal exchange and many newcomers, including me, learned to solder around this table. In the backyard, one of the admin members, MRE, experimented with his Arduino-based earthquake detection system, which he published, along with several other handy tutorials, in the book *Arduino Projects to Change the World.*[8]

More experienced and seasoned THS members operated in a peda-gogical mode of interaction with an emphasis on horizontalist instruc-tional exchanges. On one occasion, a novice in the space approached a prestigious member—who had recognized but approachable electronics expertise—with a malfunctioning USB device. The experienced hacker proceeded to ask questions about potential ways of diagnosing and fixing the device that anyone with a modicum of self-determination could try. In another instance, after I asked basic questions about the new electronic door, I was guided step by step by a member through his C program to con-trol the lock. Line by line, function by function, I followed his index finger as we scrolled down through the code he cloned to my upcycled ThinkPad. I could tell that my interlocutor was very proud of his work. For my part, I was just curious about the locking mechanism as an extension of my in-terest in mechanical locks—an ideal match that was not at all uncommon in the pedagogical and informal conviviality of the space.

Given the moral emphasis on information sharing in hackerspace cir-cuits, visitors and members came to interact through technical debates that were sometimes heated, other times playful, but always significative. These discussions often morphed into arguments, where one participant would make fun of the other for not knowing about a technical subject or object: "so . . . you do not know how TLS works?"[9] Teasing was a recurrent practice in this playful frame, but it mostly took place among perceived equals in terms of technical prowess. This banter would establish a playful match for the valuation of a technical persona. I never witnessed this kind of banter used to belittle a newcomer. There was an overall good-faith effort to welcome new people in the community. But I did witness interac-tions that marked stark gender divides: "joking," for instance, that presti-gious women-identified hackers had "bigger penises" than men-identified members with less technical expertise. Projections about the persona of gender minorities reactivated an enormous array of sedimented preju-dices that have historically plagued technoscientific circuits from which hackerspaces inherited many of their technical practices, objects, and con-stituents. It is important to note that THS was very far from being the only space where these technopolitical dynamics took place. Despite not being the dominant mode of personal interactions, they did happen there and elsewhere.[10]

In her discussion of exclusionary practices in Euro-American com-munity spaces, Sarah Davies reports the words of a feminist hackerspace founder about the matter: "The fact that women and minorities and queer people and migrants are underrepresented in the hacker and maker movement isn't indicative of there being no interest in it. The problem is not with these groups, but with the 'dominant culture' that works to exclude them."[11] It is common for newcomers without technical leanings to be left out of a circle of exchange, thus becoming "invisible." I experienced this invisibility many times for projects where I had little, I thought, to contribute. Communications scholar Christina Dunbar-Hester examined this issue extensively by situating it in connection with a much longer genealogy of power that linked performances of masculinity as mastery of technical objects—from amateur radio and electronics tinkering in the first half of the twentieth century to computer machines from the postwar period to the present.[12]

Typically, new members would persevere and learn enough to find themselves, after a couple of years, in a position to help others. Public workshops—modeled after Mitch Altman's Circuit Hacking Mondays at Noisebridge—served as occasions for inviting newcomers to pick up programming skills, experiment with coffee roasting, learn video-game design, inspect and improve electronic circuits, and tinker with all sorts of materials from fabrics to wood and spare computer parts. Through experimental practices, these common tasks would connect geographically and socioeconomically distant places, from San Francisco to Shenzhen, from Shenzhen to Tokyo. A "good member" in this circuit was a person who contributed to public events, engaged in shared projects, and despite all the intersectional blind spots and active exclusions, brought new people in. It was also common in the space to have artists working alongside Japanese and expat hackers and electronic hobbyists. On these occasions, the gender distribution would be more even; conversely, when electronics and computer topics were the focus, the place would be predominantly occupied by men-identified members (with notable exceptions).

THS projects were, to varying degrees, open to experimentation because they derived from common technologies. Conviviality was considered more important than the end result of any particular project, but that all changed when socio-environmental and political matters became life-threatening. With the onset of the nuclear disaster, THS projects started to embody the urgencies and anxieties about the collective future of Japan and nearby countries.

Learning to Perceive the Nuclear Crisis

Two weeks after the Fukushima disaster, THS members were relieved to see each other at their weekly meeting, reunited by an uncommon agenda: hacking for crisis relief. In this emergent context, hacking was widely cast as a shared skill to help in the relief and recovery process. A historic call was promptly circulated around community spaces, drawing attention to what THS members were planning to do:

> Many of our members have been cooped up in our homes waiting out the storm, but not lying idle. . . . We also have on the way several Geiger counters and Geiger tubes, from which we will be making community sensors in order to help to keep the public in harm's way informed on a minute-by-minute and hour-by-hour basis. While the initial exposure has been low, our concern is the long-term effects, food and water supply, and ground soil conditions over the next several months. Our longer-term projects include solar-cell phone-charging stations, low-energy cooking equipment, internet, wifi, and laptop loans. . . . We are calling upon Hacker Spaces all over the world and friends of Hacker Spaces and friends of friends of Hacker Spaces to help out.[13]

I came to THS once and realized that its cosmopolitan circuits were already in place. It was early afternoon on an ordinary hot and humid summer day in Tokyo. There was only one person around, one of the most experienced hardware engineers who was on a conference call with other hackers at the summer camp of the Chaos Computer Club near Berlin. At the camp were Open Hardware specialists, members of various hackerspaces, including Noisebridge, and one keen antinuclear activist who had taught himself a great deal about radiation monitoring after the Chernobyl incident. The general tone of the exchange was of mutual respect with no room for secrecy. Both sides explained the challenges in measuring and assessing the situation, bridging expertise on nuclear responses from Chernobyl to Fukushima. What resulted from the meeting (other than a powerful set of virtual ties) was a public wiki page with information about monitoring tools and techniques, drafted as soon as the THS founder hung up.

As we headed home late that afternoon, I thought about health risks and what it meant to be in Japan after the disaster. We could not properly assess, feel, or visualize the risks without specialized equipment that had

disappeared from the shelves quite literally overnight. People were pan-
icking, especially families with young children. We were afraid of our food
sources.[14] Displaced coastal communities wanted to return and reconstruct their homes and preserve their ancestral ties[15]—but an evacuation
order with a radius of twenty kilometers from the epicenter at Fukushima
Daiichi was in effect. Breaking the silence, I asked how the THS hacker felt
about his own life in the years to come. He confided that he was not concerned, since there was "nothing [he] could do to change the situation."
There was "no other place for [him] to go." Then he continued in much
more reassuring tone: "If you hang out with us, you will learn a lot about
radiation." And so I did—as if the research topic had chosen me and not the
other way around.

After March 11, 2011, being present at THS meant being surrounded by
radiation-monitoring volunteers and participating, even if as just a listener
initially, in debates about monitoring techniques and design challenges
for assembling open instruments. Engineers and hackers were eager to
discuss for countless hours what they could do in the face of the unfolding
crisis. As an anthropologist learning the ropes, I took part in some of the
simpler technical tasks, such as soldering circuit boards for the first prototypes. I also volunteered for data collection and attended various iterations of the radiation-monitoring workshop dedicated to the assemblage
of (open) Geiger counters. First developed at the turn of the twentieth century by the physicists Hans Geiger and Ernest Rutherford, Geiger counters
detect the presence of radioactive particles by converting them into electric pulses that are counted as "events."[16] The higher the count, the higher
the risk of contamination for human and non-human lives.

Learning to perceive the crisis as it unfolded required building a
common circuitry through which volunteers found the conditions to
teach themselves and others about radiation. Given the limited capacity
and availability of people and tools at the time, it was difficult but important to secure a seat at the radiation-monitoring workshops held by
THS hackers who, despite not being nuclear physicists, had enough background to distill a basic course. To sustain their experimental work, they
asked participants to cover the costs of their Open Hardware–based Geiger
counter prototype. Their workshops emulated the convivial format we
found in other hackerspaces, but they had a critical, if not somber addition: they started with a crash course on the nature of ionizing radiation
and its different kinds (alpha, beta, and gamma), the type of monitoring

they could reliably perform, and what was known about environmental migration patterns, half-life, and dangerous levels of contamination for the most frequent radionuclides released from the nuclear meltdown (such cesium-134, cesium-137, and iodine-131).[17]

A serious obsession with monitoring equipment was brewing at the time. Some THS members were even bothered by the fact that Geiger counters took precedence over other community projects and topics of discussion. Concerns about personal health and levels of exposure, curiously enough, took a backseat to the imperative of advancing the state of the art in collecting, mapping, and visualizing radiation data. The rationale for those deeply committed to radiation monitoring was simple: safety was a challenging variable because it depended on identifying radiation hotspots to avoid, but how could hotspots be identified without proper instrumentation and monitoring? With escalating tension around damning reports on the problem of "regulatory capture" by the Japanese nuclear industry, it was difficult to find guidance due to high levels of governmental opacity.[18] Historian Jeffrey Kingston identified the source of this problem in the so-called nuclear village, a term colloquially used to describe the reprehensible "collaboration" between the Japanese government and the nuclear industry, but also to refer to the actual places where new corporate employees were sent to be indoctrinated in the purported benefits of nuclear energy as "cheaper, cleaner, and more efficient."[19]

By following the group of hackers turned radiation vigilantes and attending to the intricate details of radiological measurement, I started to slowly and ethnographically describe relations I observed online and offline.[20] As part of the work of sociotechnical description, I populated network maps depicting ties between technologies, technologists, and institutions. After completion, I used these maps to analyze the transformation of THS from a convivial hobbyist space into a key site for assembling serious radiation-monitoring tools.

Assembling Common Tools

A couple of years before the disaster, THS already had a fair share of projects in common with the nascent Open Hardware community. Participating in the space, after all, were itinerant hackers in addition to the stable membership of local computer hobbyists, artists, and senior engineers working for the global IT industry in expert domains such as information security and network administration. From brainstorming sessions, a wide range of projects for crisis relief quickly emerged right after the nuclear meltdown. Common tools and accessories were designed and developed to completion, such as flexible battery chargers, solar panel applications, seismic sensors, and a detailed wiki page with scientific reports on the nuclear crisis.

Out of an innocent interest in using microcontrollers to automate gardening tasks came the clever idea for the Kimono Lantern, which was initially meant to be a simple solar-powered light for a restaurant. The project was rather simple: it consisted of a solar-powered circuit mounted on the metal lid of a glass container. Its technical ensemble included an LED, a solar cell, a rechargeable battery, and a light sensor. During the day, the LED would be off, while the system collected energy through its solar cell to charge its battery. With sunset, the charging process would be stopped and the LED would turn on to illuminate its surroundings. It was a beautiful minimalist design by one of the THS resident hackers, Chris "Akiba" Wang, nicknamed after Tokyo's dream "electric" district of Akihabara.

After March 11, the Kimono Lantern became one of the first projects to be offered to the communities affected by the tsunami in northern Honshu. Soon after the hackerspace network's wide call for help, THS became a critical site for bringing local and foreign volunteers together. Parallel networks were rapidly mobilized, both online and offline, to discuss possible technical ways to help assess the escalating nuclear crisis. A small group of designers and engineers from Japan, Germany, the United Kingdom, and the United States, for instance, was formed around an internet service platform called Pachube. Its goal was to collect, localize, and distribute data online from personal monitoring devices. Another key grassroots initiative was led by an influential Japanese Free Software developer with a group of volunteers from the Open Street Map project, who rapidly set up a crisis information platform called Sinsai.info. Aiming to collect and distribute geotagged information about relief efforts, they helped coordinate tasks of local crisis response teams. To borrow the expression of political theorists Antonio Negri and Michael Hardt, social technologies of the common[21] were readily made available in these trying and confusing times.

In metropolitan Tokyo, a group of internet entrepreneurs was quick to mobilize and discuss the use of open technologies to measure radiation levels. Thanks to their prestige and wide network of contacts, they gathered an expert group of volunteers around the Next Context Conference, which served as a catalyst to materialize the collaboration between hardware engineers and software developers with an initiative across the Pacific named RTDN.org, a web service for aggregating and distributing geotagged data. At the conference, the network RTDN.org was renamed Safecast after the suggestion and offer of a domain name from former chief software architect at Microsoft, Ray Ozzie.

From the outset, the Safecast project was adopted under the umbrella of Momoko Ito Foundation, a nonprofit dedicated to the research of cultural exchange between the United States and Japan. The foundation served as the legal counterpart of Safecast, handling donations and the project's trademark. This institutional support was secured by the former MIT Media Lab director Joi Ito. Unlike other independent radiation mapping efforts, such as Geigermaps JP and Japangeiger, Safecast managed to secure funding and donations, partnerships with educational institutions, and a volunteer network on a truly international scale. In thirty days, the project formed a core team of expert technologists who would develop (in their spare time) some of the key technologies and define a replicable method for collecting, geolocating, storing, and distributing radiation data.

There were no previous work ties among most of the technical volunteers of THS and Safecast who hacked Geiger counter devices in less than a week of cooperative work. What they shared instead was the familiarity with common tools and frameworks and the capacity to harness considerable contributions through hackerspace-based ties. As Akiba characterized their serious independent achievements: "It is time now for us to show the commercial IT guys what a bunch of dirty hackers can do!"

Despite concentrated efforts to foster public participation in collective projects in the aftermath of the disaster, most independent hardware projects in Japan did not succeed in creating a community, since they were limited to small-scale distribution of affordable Geiger counters. By contrast, Safecast grew in public participation and in the number of partnerships and delivered tangible results in a matter of months after the disaster, attracting attention from the media and securing funding from a wide range of domestic and international sources. Volunteers could be counted in the order of hundreds, including partner associations in Japan and abroad. Funding from Softbank, for example, was injected into a stationary sensor network project built in partnership with Tokyo Hacker

Space and Keio University. Financial support also came from four crowd-funding initiatives: two Kickstarter and two Global Giving campaigns. In August 2011, Safecast was invited to a prestigious art exhibition, "Sensing Places," organized by Ars Electronica, where its Geiger counters were displayed as design pieces. In September 2011, the Knight Foundation offered a prestigious grant to the group in recognition for the journalistic qualities of its radiation mapping effort.

Safecast became too big too fast for Tokyo Hacker Space. Eventually, it had to move its temporary headquarters out of THS, where it had benefited early on from intensive brainstorming sessions, hardware prototyping, and wide circulation of itinerant hackers. As the Safecast project and its dataset grew over time, an ample office space was offered by the Creative Commons–inspired design firm Loftwork in Shibuya, Tokyo.

Mutual Aid for Serious Bricolage

In an internal debate among THS members regarding how to position themselves as experts without scientific credentials, one member shared his discontent with the data released by the Japanese government, arguing that "anything but full disclosure of [radiation] data and sources [were] totally unexceptionable. Just because you don't understand the data that doesn't mean someone else can't." His rant was followed by a humorous message from a charismatic THS hacker, paraphrasing an anecdote that had been popularized by early AI researchers Hal Abelson, Jerry Sussman, and Julie Sussman: "This reminds me of this Zen koan written by Hakuin Ekaku in 1750. A young monk asked his master, 'If the source code is freely available but nobody reads it, is it still open source?' The master laughed and walked off. The monk was enlightened."[22]

Both Tokyo Hacker Space as a self-managed convivial laboratory and Safecast as a spin-off project of technologists and internet entrepreneurs can be described as sites for the practice of hacking as serious bricolage. Following anthropologist of science Sharon Traweek's methodological orientation for the study of group histories through the development histories of their instruments of knowledge,[23] we can learn about hackers as bricoleurs from the development process of their open instruments.

To illustrate the operational logic of mythical thought in his book *The Savage Mind*, anthropologist Claude Lévi-Strauss introduced the figure of the bricoleur—a French expression for a jack-of-all-trades who uses whatever materials are at hand to fix or repurpose technical artifacts. In this metaphorical sense,[24] bricolage stands for a distinctive kind of knowledge acquired through a hands-on approach without a clear definition of a project. It prioritizes secondhand raw materials and roundabout, patchy ways over formalized procedures. According to Lévi-Strauss, "the poetry of bricolage . . . comes from the fact that it does not limit itself to accomplishment or execution; it 'speaks,' not only with things . . . but also by means of things."[25] The bricoleur, he insists, creates facts by breaching conventions and changing the state of the world, whereas the engineer uses facts to create structures. As a poetic opening to the world, the practice of bricolage "decompose[s] and recompose[s] sets of events (on the psychic, sociohistorical, or technical plane)."[26]

Surpassing the distinction between the bricoleur and the engineer, science studies scholars have demonstrated the ways in which bricolage is also constitutive of technoscientific practices.[27] Social psychologist Sherry Turkle and computer scientist Seymour Papert's work has the merit of being the first to highlight the bricoleur approach in the field of computing. They examined "programming styles" to illustrate the interplay between concrete and abstract reasoning in what they call "epistemological pluralism" in computing.[28] For our purposes here, this line of research is key to helping us understand how learning to learn takes place informally at hackerspaces through open technical-object-oriented sociality: "There is a culture of computer virtuosos, the hacker culture," Turkle and Papert declare, "that would recognize many elements of the bricolage style as their own."[29]

Collaborative work around open technologies in Japan post-3/11 revealed itself to be a form of serious bricolage for crisis response. As soon as the nuclear crisis became mainstream news, radiation measurement devices became hard to find. Shortages of Geiger counters and Geiger-Müller tubes in Japan and abroad were felt rapidly as citizens searched for alternatives in the face of government denial and conspicuous secrecy. University experts were quoted in popular commercial media outlets asking the population to leave expert systems to technoscientific experts. The message was straightforward: data from cheap Geiger counters was not reliable, and untrained laypersons should refrain from spreading unreliable data. Experience once again tells us that it takes only one authority to deny access to a certain resource to a group of hackers for that very resource to become available in roundabout ways.

Hackers as unlikely actors came to the fore with technical responses to the disaster that were articulated locally and internationally. Japanese, Spanish, Chinese, and North American hardware designers and small vendors offered their own Geiger counters on the internet based on Free and Open Source prototyping tools (figure 5.1). Engineers from Tokyo posted regular information about the development process in blogs, creating an interface for volunteers to be included in the fold of experimental work. Moving beyond the hobbyist circuit, community-led projects quickly attracted commercial and public attention with the outbreak of the disaster, leading to the conversion of modest open technology gifts into products for the niche of portable Geiger counters—such as the example of Open Geiger and its commercial distribution in Japan by Atarashimon, Inc.

In parallel with small-scale hobbyist experiments and commercial ventures, Safecast bricoleurs had the brilliant idea of placing their open Geiger counter inside a bento box and baptizing it "bGeigie." By integrating available technologies at hand, teams of skilled volunteers created a portable device with sufficient autonomy and reliability to perform measurements while its user was walking, biking, or driving across Japan. The mobile approach to radiation mapping was initially tested with a netbook attached to a GPS device and a commercial Geiger counter. This technique was inspired by a community practice of wireless hacking called "war-driving" in which hackers survey a particular region with a computer equipped with

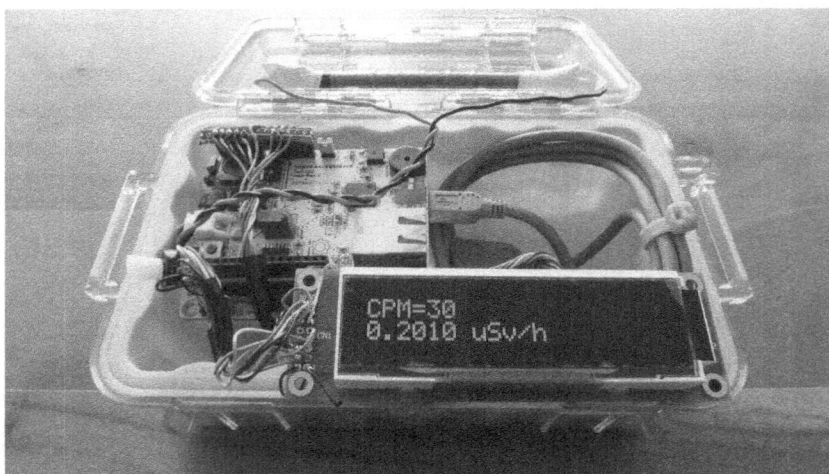

FIGURE 5.1   Early Geiger counter prototype, Akiba's NetRAD

a wireless network device hooked up to an external antenna. The practice of "safecasting," or "radiation war-driving," proved to be not only feasible but also very convenient. When the bGeigie was attached to a car that volunteers drove around Japan, radiation data started to be collected for the project, properly located by the integrated GPS module and stored on an SD card for later publication on the internet. This experimental bricolage took time. Before obtaining a positive result in their mobile approach, Safecast volunteers went through countless hours of trial and error, teaching themselves and others through hands-on experimentation with prototyping tools for environmental monitoring.

The serious bricoleur work for the bGeigie eventually stood out as an achievement of integration and multiple field expertise coordination. Working with the informational resources at hand, hardware hackers assembled the device from gifts of various open technology projects. The output of their open Geiger counter—log files with device identification number, GPS coordinates, gamma radiation levels, and time stamp—would be shared by Safecast as counter-gifts in the form of public domain data. By releasing radiation data under a Creative Commons zero (CC-0) license, no restrictions were imposed for reading, copying, modifying, or even appropriating the data without attributing the source. This orientation to gift giving among strangers came from one of the board members who happened to also be on the board of directors of Creative Commons and various internet companies. Commercial restrictions on open licenses were perceived to be detrimental to the efforts of network building with for-profit and nonprofit organizations—except, of course, when the arrow of appropriation was reversed, going from the digital commons to digital products based on "permissive" open licenses (in contrast to "reciprocal" ones) that allow for shutting off open technology circuits.[30]

After circulating its early network-wide call for help, THS received two old Geiger counters from a surplus-outlet shop of scientific and industrial grade equipment called Reuseum. Driven by the same orientation for serious bricolage, on May 1, 2011, the first THS prototype for georeferenced data logging was announced. Shortly after the online publication of the fabrication process—step by step, with pictures and an explanation of the technical challenges—the device was made open by publishing its circuit schematic and firmware code. The development of the NetRAD extension board was done by the same skilled engineer who had hacked the old military Geiger counters, Akiba, and who happened to have a pick-and-place machine to populate and solder very small surface-mounted components onto circuit boards, thus speeding up the process of prototyping considerably. When he first approached a company to buy the industrial machine and install it in his apartment, Akiba recalls being laughed at. His apartment in Tokyo, THS members would remind me, was on an upper floor and there was no elevator.

The first prototypes integrated an external, professional-grade Geiger counter, International Medcom Inspector, donated to Safecast shortly after the Fukushima Daiichi accident. One of the founders of Medcom, Dan Sythe, joined Safecast in its early stages to share the radiation-monitoring expertise he had cultivated since the Three Mile Island nuclear accident in 1979. Sythe offered much-needed advice on calibration and measurement techniques, helping to evaluate Safecast's prototypes along the way. The collaboration between the vetted expert and the lay-expert hackers was actualized on the basis of a new kind of reciprocity: while Sythe helped to legitimize Safecast with his industrial-grade sensors, his company benefited from the commercialization of open designs by distributing kits and preassembled versions of Safecast's bGeigie.

Software code for the bGeigie was created by branching public repositories of common code from Japanese and international initiatives in hardware prototyping. Software for an Arduino-based Ethernet extension board, for example, came early on from a crowd-mapping initiative developed by a prestigious hardware designer from Japan, who rapidly became a respected contributor among THS and Safecast volunteers. Akiba recalled reading his code for the first time and being delighted with his organization, cleanness, and elegance. THS members had to meet the extraordinary designer, so they invited him to come to the space to give a presentation.

Hackers-as-Bricoleurs and Their Others

In their experiments with bricolage, Safecast volunteers did not intend to displace other radiation mapping initiatives, far from it, but to collect, organize, and publish data through what they assumed to be the best possible way to mitigate the problems of unavailability and unreliability. Safecast members tended to describe their position as being a "mediator" between the government, the scientific experts, and the public. In doing so, they often compared their work with that of other projects and criticized governmental and private sector initiatives for their inefficiency, technical opacity, and lack of procedural transparency.

One of the most important governmental attempts at radiation mapping and longitudinal monitoring was implemented by the Ministry of Education, Culture, Sports, Science and Technology (MEXT). Public outcry and heavy criticism was directed toward the aerial radiation monitoring performed by MEXT, which started in January 2012. The public was skeptical of the low readings published in the governmental report and questioned the sampling method as an attempt to cover up the situation. Another initiative came out of a partnership between Kyoto University and National Instruments. The project Kurama consisted of a mobile system for radiation monitoring installed in public transportation vehicles of the Fukushima prefecture.[31] Kurama closely resembled Safecast in its mobile approach, but it proved to be too costly, nonreplicable, and closed to public scrutiny—hence, the view that it was just another ineffective government-sponsored partnership.

It was not long before the black boxes of the MEXT monitoring stations were opened. Manufacturing and installation of monitors were arranged by MEXT with private contractors such as NEC, Aloka, Fuji Electric, and Rhinotech in order to deploy a large number of solar-powered stations across Japan, with the largest concentration in Fukushima. The stationary sensors were meant to provide localized and longitudinal data displayed in navigable maps on the website of the Nuclear Regulation Authority. In a typical hacker approach, MEXT stations were eventually examined in detail by two Safecast volunteers. While performing a "Safecast drive," a core member found a malfunctioning station with an open compartment. He could not help it—he had to stop. By approaching the station, investigating its internal parts, and taking photos so that he could share his finding with others, he realized that it was not weatherproof, which explained why the black boxes failed so often. Data provided by the monitoring stations have also been criticized by local resident and activist groups, such as the Association for Citizens and Scientists Concerned about Internal Radiation Exposures and Greenpeace. The alignment of critical positions in respect to governmental data—as "manipulated data" to convey smaller figures for affected areas—was temporary, however. Hardware hacking was only momentarily aligned with other critical voices post-3/11 because the consensual position within Safecast was to avoid engaging in public debates regarding nuclear power. This controversial position had its own history within the group.

Responsibility or Distributed Response-ability?

A decade after the outbreak of the disaster, long-term exposure to (harmful) radiation began to gain much more attention. Well-resourced and established, Safecast gained international recognition for the most comprehensive dataset ever produced using a collective approach to radiation monitoring. The Fukushima prefecture started using Safecast as its source in a major victory for the project. Wherever "citizen science" was called into public debate, Safecast was given as an example of what political economist Yochai Benkler called "peer mutualism."[32] Safecast's Geiger counter prototype reached maturity and stability, becoming a pedagogical kit to workshop with kids and adults. Through partnerships the project established with scientific instrument makers, such as International Medcom, the kit was converted into preassembled commercial products. Its volunteers moved on to experiment with other forms of monitoring, such as air quality, only to encounter similar challenges regarding sensing and sensors.

Safecast started to circulate widely and internationally as a symbol, standing for the power of self-organized groups in responding to the public and environmental health disaster in Japan. As a disputed technical and scientific symbol, however, it spoke much more about the position of those who were mobilizing it than about the commoning practices for assembling its base technologies: a disconnect all too familiar to hackers.

In the broader context of community-based responses to the triple disaster, "openness" became a key technical practice for commoning within serious limits. The importance of Free and Open Source bricolage for crisis response became identified with the capacity of lay expert groups to shift power relations with more comprehensive and reliable data. More than a demand for individuals to take action, "response-abilities" were substantiated through shared work around common technologies—with material, scientific, and ethical entanglements, studied by feminist technoscience scholar Donna Haraway,[33] that characterized the urgent shift away from the liberal understanding of atomized responsibility. The collective effort integrated in common circuits, spanning local and cosmopolitan organizations and independent hackerspaces, led after all to unprecedented sociotechnical results. Tokyo Hacker Space and Safecast's technopolitical mobilization ended up transforming but not at all dissolving the power dynamics between credentialed experts, lay experts, and volunteers by actively crossing the boundaries of what cyborg anthropologists have called the "technoscientific citadel"[34]—that is, the locus of technoscientific power for determining who gets to speak the truth about the risk of (harmful) radiation exposure.

It is important to emphasize that THS and Safecast were small pieces of a much larger puzzle of community responses to the Fukushima disaster. Sociologist Aya Kimura studied, for example, community initiatives on radiation testing of the food supply, making an important intervention in the debate about gender, care, and response-ability.[35] Kimura demonstrated that in boundary-making practices that separated emergent community initiatives from antinuclear groups, gender played a fundamental role in shaping local organizations' contributions to crisis response. The neighborhood-based groups that Kimura studied were spearheaded by Japanese mothers who were deeply concerned about the health of their families, leading them to take on the task of independent food monitoring. Through their own familial and neighborhood-based circuits, they created parallel ways for commoning that looked quite distinct from that of THS and Safecast. However, they all involved commoning techniques to share, learn, and teach as much as they could.

In his ethnographic work on morality and ethics, anthropologist Jarrett Zigon also examined the limits of the liberal understanding of "responsibility" with its individualistic frame.[36] In parallel with Kimura, Zigon proposes an alternative that could be better described in terms of response-ability: a relational ethics that is grounded in an ethics of attunement to the other. The point of departure of relational ethics, Zigon suggests, is the question "How is it between us?" Response-ability depends not only on intersubjective demands that we place on each other but also on the politico-economic (and we would like to add technoscientific) orders that shape relations of command and control to constrain our capacity to attune and respond.[37] It is in this particular sense that hacker responses to 3/11 can be examined with a much sharper focus not only as an expression of the hubris of engineers as Promethean, neoliberal pawns but as a collective response-ability that was made possible through a very specific common circuitry. Hackers as mothers, as well as mothers and hackers, all had different response-abilities in the aftermath of the disaster.

The Fukushima crisis has been an object of a wide range of citizen, academic, and journalistic accounts, forming a public sphere with multiple perspectives in dispute. In face of the unfolding multi-scalar socio-environmental disaster, Safecast carved its own space as a prestigious project. Its volunteer work, broadly conceived, consisted in not feeding the heated debate on natural versus anthropogenic classifications of the disaster, which became a central albeit false binary after it appeared in the report of the Nuclear Accident Independent Investigation Commission appointed by the Japanese government.[38] Instead, Safecast volunteers helped establish a distinction between information and evaluation, drawing a new symbolic division between the goals of the independent project versus the role of credentialed experts. By examining the constitutive circuits of the project alongside its public manifestations, it is possible for us to identify their efforts in rendering the new boundaries between nuclear experts and nonexperts visible. "We're actually working very closely with Keio University at the moment and a number of their members are going with us on the data collection drives," said one of the most active volunteers in an early post to the whole team. "We have already been discussing with them about data collection procedures and how to make them scientifically credible. They are also helping to establish a facility to handle checking and calibrating equipment that we'll be distributing. This will be done by radiological experts within Keio University but I'm hoping to document their procedures and make them transparent."[39]

The technoscientific orientation of THS and Safecast hackers did not emerge, curiously enough, from the polarized debate of antinuclear and pronuclear organizations post-3/11. Contrary to a likely assumption, Safecast hackers did not subscribe to the antinuclear movement but engaged in establishing their practices of "hacking" as a form of serious bricolage. Their project did not stem from the genealogy of antinuclear hacktivism either. In that genealogy,[40] we find initiatives such as Netstrike.it (an Italian activist initiative against military nuclear programs in Europe) and Worms Against Nuclear Killers (also known as WANK, an anti-nuclear prank by Australian hackers that infected NASA's Space Physics Analysis Network, protesting the potential release of plutonium from the Galileo space probe).[41]

Safecast's "other" was the official governmental response to the nuclear crisis, but it did not take the path of direct-action hacktivism. Despite growing antinuclear sentiment and mobilization in Japan, where most of the population did not want to restart halted nuclear power plants, it was not by partaking in protests that the project captivated and enrolled new volunteers. "All the core people [at Safecast] are very diligent about trying to stay objective and neutral," one of the founders once observed, "and not trying to take a political side on any of the many political issues that are around this, both on food contamination and the appropriate usage of nuclear energy. . . . We have a strong shared vision, which is to gather the data, to share the data, to promote Open Source, Open Data. More data, more people sharing it, it will alleviate ignorance and anxiety. That is what is common among people in Safecast."

Early on, hackers and non-hackers would embrace the idea of being "pro data" instead of being anti-nuclear or pronuclear, bringing to the fore a new position in the debate regarding our shared radioactive condition. As the commoning logics expanded beyond software development practices, volunteers faced new sociotechnical challenges that demanded a reconfiguration of openness in concrete ways. Open Hardware projects, for example, dealt with serious issues of competition for electronic components and parts, as well as with the question of scale in production and distribution. Not only the materiality of the digital commons stood in the way, but the scarcity of multi-field expertise as well—as most projects bridged analog and digital electronics, 3-D modeling, data analytics, and knowledge of communication protocols, among several other areas of specialized knowledge. For Open Data, the question was not only about licensing but about stewardship, trust, and response-ability. Open Data initiatives frequently face problems that add up to the mix of infrastructural issues; it is not only labor-intensive and time-consuming to make information systems well documented and therefore accessible. The generation, organization, and interpretation of data in heterogeneous information systems lacking in interoperability present a challenge. These problems are compounded with the serious issue of data stewardship with equitable distribution of expertise and power to make meaningful use of the data for community purposes.

All of these fraught technopolitical issues attest to the limits of "openness" as a mode of data governance, which is hampered by long-standing exclusionary forms of training in science and technology. The question of openness is also directly related to the open issue of response-ability. As an anonymous contributor once asked the team of Safecast volunteers: "If Safecast is going to be seen as having an authoritative teacher's voice, what will underpin that authority?" One can always look for credentialed institutions to encounter the modernist demand for centralized authority, but hackers tend to look within their own self-organized communities for sources of legitimacy. We could hear from its members that Safecast did not want to be involved in societal decisions, echoing a long-held division between technical and political work within the collective—that dangerous but formative orientation in science and engineering. As we saw, this controversial position had its own history. It started early on with a message thread between two core members over the prickly question of data interpretation: "Is Safecast going to help people interpret data or only take up the role of data provider?"[42]

The defense of "neutrality" among Safecast volunteers with respect to data interpretation eventually won over both internal and external demands for the project to take a public stance. Their elaboration on what it meant to be "pro-data" in the context of government secrecy had a strong hold. During a public workshop in Tokyo, as a technical volunteer explained the characteristics of the DIY radiation monitors to an academic nuclear physicist, the credentialed expert clearly grew increasingly uncomfortable and skeptical, contradicting the volunteer's explanation at every opportunity. The latent conflict broke when the volunteer finally conceded: "We decided to leave interpretation to specialists. We just collect data to be analyzed by them." This encounter brought to the fore recurrent practices of boundary-making between open technology practitioners and scientific experts in the context of crisis response. An invisible wall between areas of expertise emerged as a reaction to criticism from field specialists, pointing to more established forms of credentialed authority and legitimacy of those who get to interpret the data.

The question of safety and risk was recurrent in the Safecast public mailing list, but it was considered a hard one so the core project members constantly avoided it. In several public interventions, Safecast members digressed from the topic of risk, trying instead to offer information on how to carefully evaluate different forms of instrumentation and measurement. In a controversial post, one of the members (a computer security expert by day and an active volunteer by night) addressed the problem of risk by exposing the specificities to be taken into account. His statement was posted only a month after the project had officially been launched: "Again we want to make it clear that we are not radiation experts nor health physicists—we aren't making any claim about how safe, or not, any of these measurements might be—rather, we are trying to find and provide data that could be important for residents of these areas so that they can make informed decisions on their own."[43]

In the context of open technology projects such as Safecast, expertise is performed within a common circuitry that authorizes and legitimizes volunteers, depending on the perceived quality of their gift contributions. Drawing the contentious but symbolically efficacious line between technology experts, antinuclear activists, and nuclear scientists helped Safecast to avoid engaging in seemingly intractable scientific controversies. While Safecast members worked to position themselves as trustworthy technoscientific agents, gaining respect from the private and public sectors in Japan and abroad, the project proceeded to guard its borders by inviting skilled engineers who were not antinuclear activists to join the core group, therefore moderating public participation. In this regard, Safecast hackers had a double relationship with the discursive function of expertise and the performative dimension of technical knowledge: their practices relied on established institutions as much as they criticized those and distanced themselves in an attempt to create conditions for more autonomy and openness; a safe distance was, overall, maintained to strike a balance between technical mastery and improvisation (grounded in several years of cultivation in electronics, computing, and engineering of its volunteers). As a symbol for collective identification, Safecast functioned by breaking away from the institutional practice of nuclear science, while still operating on its edges, accessing grants and taking advantage of prestigious research centers, using its language, standards, and procedures and establishing productive partnerships for gift exchange, free from the career-building constraints and disciplinary practices of academia. In this process, response-ability revealed itself as a fruitful analytic, but one that has yet to become a concern among hackers as they champion openness, at once, as a necessary point of departure, a desired outcome, and a technopolitical practice for mutual aid.

SIX

Gniibe

Hacking to Do No Harm

We were on our way to Mount Akagi in the Maebashi region on the island of Honshu, Japan. My partner in technical pilgrimages, Niibe "Gniibe" Yutaka, sat quietly next to his mother, who was driving us past a "singing road." As we slowly climbed, so did the microsieverts per hour on the small seven-segment display of my open Geiger counter, firmly attached to the car window inside a waterproof acrylic box—a creation of serious bricolage. I saw the increase in radiation levels and promptly communicated it to my hosts in the front seats, but they were unfazed.

I was on a mission to collect radiation data and contribute it to Safecast, but also to continue the trajectory interviews with Gniibe. A Tokyo Hacker Space (THS) member let me borrow a spare monitoring kit to take on my trip to Gniibe's home town. In the kit, I found copies of a business card with contact info for the Safecast project but with the generic inscription "Volunteer." In between interviewing sessions in Maebashi, Gniibe and I took quick strolls around town, had *yakimanju* and tea, and soaked in the nearest *onsen*. The raw material of ethnographic research is made not only of barriers and challenges (linguistic, technical, socioeconomic, interpersonal, and spiritual) but also of unanticipated communion through ludic and idle moments, such as the ones I had with Gniibe, Mitch, and Nala. Some of us would argue (and it has been an ongoing debate in anthropology since the 1980s) that these are the moments that actually have weight in anthropological encounters. In any case, our pauses were necessary points of refuge in the profusion of topics, memories, and insightful tangents we would inevitably take through free association.

I had first met Gniibe in Tokyo during an unbearably hot summer in 2010, right before a meeting of the Free Software Initiative Japan. We had no idea what was about to mobilize us all with the triple disaster. Since then, we have met on several occasions to record interviews but also to entertain long conversations about what we decided to call amongst ourselves, for lack of a better term, "computing and ethics."

Arriving at Mount Akagi, Gniibe called my attention to the fact that Lake Onuma had been found to be contaminated with high levels of cesium-137, and he expressed his deep disappointment with the situation. *The local government recently passed a fishing ban.* I was overwhelmed by the beauty of the place, but I felt I could not allow myself to appreciate it. I wanted to join him and his mother to "safecast" it, but that was not the main purpose of our trip. I did not want to ask too much of their kindness. His mother had come to worship at the local temple, after all. Akagisan occupies a very important place in the local cosmology: its inactive volcano is a site of worship that figures in popular legends (*densetsu*) in the image of a great centipede mountain goddess. Its pristine waters were said to have healing properties until now. Gniibe told me his mother was well-connected in the temple through her poetry group. Once her religious observances were completed, it was time for us to return. I felt I had done a minor service myself, but not the spiritual kind. I had collected a small amount of data to contribute to the common pool as an anonymous volunteer. At the time, we did not know if the problem would be perennial for Lake Onuma. Nuclear scientist Eichii Suetomi and his colleagues from University of Tsukuba later published a paper with an updated model that predicted that low levels of cesium-137 would persist in the lake for at least another thirty years.[1] This paper made the scientific and mainstream news very quickly. Faced with the potential of invisible contamination, we were all experiencing the place in a melancholic way.

Upon my return to Tokyo, I proceeded to upload the data through a web service Safecast volunteers had implemented. In an instant, I was presented with a map representing geolocated radiation levels with scattered dots. It mapped with "good enough" accuracy and precision where we had been with a color-coded scale—a scale that had its own history of contention over how to represent dangerous levels of contamination.[2] On our drive, we had collected 3,453 measurements, represented by a string in a text file with the identification number of the device and version of the hardware firmware I carried, measurement of the radiation dosage, GPS coordinates, and time stamp. Our data points displayed 0.15 micro-sieverts per hour on average, considered a normal level of naturally occurring radiation.[3]

We started to slowly understand the extent to which people struggled across the country to make sense of their radioactive condition. The more we learned about the situation, the more we collectively felt that non-expert data collection was justified, especially given the dismal governmental efforts to measure and inform the population about radioactive contamination. The popular characterization of Promethean computer technologists who had to fashion themselves as independent technopolitical agents was a very poor explanation for what was going on through mutual aid in Japan. Safecast was inevitably caught up in numerous political disputes in this context—from imported libertarian-technologist stances, which eclipsed the social, to local antinuclear sentiments, which obliterated the technical in its moral quandaries. There was no way to avoid the centripetal force of the disaster, pulling us into a cauldron of technopolitical memories and practices. Everyone was getting cooked in pronuclear and antinuclear narratives, even those who thought the divide was pointless. There was no way to avoid being placed in one side of the fence (of the nuclear facility) or the other. While some were far more affected by the disaster than others, everyone was differentially implicated.

Back in Tokyo, things looked quite different than they did in Fukushima or Gunma or on Twitter. Contrary to my expectation, Gniibe was not at all enthusiastic about radiation mapping. Generating data was important, but more important in his opinion was to create conditions for responsible interpretation of the results by people without scientific training. Instead of doing it himself, given his technical expertise, Gniibe relied primarily on the government of his prefecture to conduct measurements of his children's school, even though he did not put much trust in the government's data overall. Despite being perfectly capable of helping with the independent monitoring efforts that I brought to his attention, he could not get himself to agree with the way Safecast volunteers presented the data. *If their movement is for education on how to build Geiger counters, or build more Geiger counters, I could support the idea, but if they exercise their power to mea-sure radiation and just upload the raw data and distribute it without interpreta-tion, it is kind of irresponsible.*

Time accumulated memories with the unusual acceleration that is typ-ical of a hackerspace: I had traversed the common circuit of hackerspaces in Tokyo, San Francisco, and Shenzhen and found myself now in Hong Kong. Somewhat sheltered from the bubbling streets of Wan Chai, Gniibe and I sat at a restaurant after a long day at the local Open Source confer-ence to do what we had been doing for the past two years: meet with no time constraints to discuss his trajectory. I kept asking about his personal opinion on various technical matters when he interrupted me, rather abruptly for the very first time. There was "too much I" in my questions, he said, expressing what anthropologist Dorinne Kondo has identified in her studies of self-making as the relational "I" in Japanese language, used to mark hierarchy, intimacy, distance, and contextuality.[4] Gniibe pro-ceeded to explain what he meant by writing in Japanese the expression "I only sufficiently know or realize" on the back of a piece of paper where he printed his Gnu Privacy Guard (GPG) fingerprint for distribution at the conference.

Given his impressive contributions to various subsystems of the Linux kernel at a time when he was one of the very few Japanese developers in contact with the transnational community, Gniibe had been invited to give a talk at the Open Source Lab meeting in Tokyo in the early 2000s. Disappointed with the unbalanced focus of the talks on technical matters rather than the moral aspects of software sharing, he decided to prepare an entirely different kind of talk—on how to dig wells. It was an awkward moment for the audience as Gniibe digressed into the story of how he and his brother, who never got along, managed to dig a six-meter-deep well with minimal communication. *I explained that sometimes it seems very diffi-cult to communicate with engineers from abroad or engineers with a completely different background, business, or culture. See my example. At that time, I had completely given up communicating with my older brother, but it was actually possible as long as one has enough passion, a clear goal, and some sort of model. If communication is not that good, it is still possible to cooperate.* This was the lesson he wanted to convey to the audience with respect to his participa-tion in internet-based software development without being explicit. *Most of the time, whenever I am invited to Asian countries to talk about computer tech-*

nology, they do not ask about the history of Free Software or its practice. They care about technology and skill. I think morality is as important as skill.

Niibe Yutaka is considered a willful veteran among Free Software enthusiasts, known for his strong opinions on matters of "software freedom." Possessor of an iron discipline and discreet generosity, he has contributed substantial pieces of software to the commons—so substantial that they are running on billions of devices around the planet. At home and abroad, he goes by the handle of "Gniibe," a nickname he adopted for its nod to GNU, directly identifying his persona with that of other Free Software activists. Despite sharing a common identity as a "hacker," he does not understand political action in the terms of his Chinese or Euro-American peers. Whenever asked to elaborate on the issue of freedom (*jiyuu*), he prefaces that his position is unusual in comparison to commonly held arguments in the community. One illustrative case comes from his engineering and programming practices—that is, his Free Software and Open Hardware work, which he does not regard as exercises of personal freedom but as active and conscious efforts of preventing himself from causing harm to others with his technical competency.

Gniibe's maternal grandfather was a small factory owner, known for his engineering competence in building and running knitting machines. His father is retired from the Japanese Air Force, and his mother held various clerical jobs, including being a salesperson for a cosmetics company. Gniibe grew up in a middle-class family without economic hardship. As far back as he can remember, his upbringing was marked by his parents' avoiding any form of conflict with relatives. *In my childhood, it was somehow prohibited to talk or discuss topics on the news, especially politics. My father and my mother never disclosed who they voted for.* He argues that it is important to continue to practice this approach with his own children to keep a peaceful home (*ie*). He would openly disclose preferences related to his hobbies, such as baseball, sumo, and judo, or his favorite chain restaurant. *I support the Yakult Swallows baseball team; I am a fan. And I do not like the Yomiuri Giants. Yomiuri is questionable. They use their power of mass media to control the public.*

His informal training started at a very young age, when his mother gave him a soldering iron. He was eleven years old. Around the same time, he managed to convince his parents to buy a mail-order course on electronics for hobbyists. *Perhaps it was my older brother who started the course, but he got bored with it, and I continued. My memory is clear about the personal* computer, though. *It was my brother who insisted on it. He said it was very important for my parents to buy a personal computer because it was a device of the future. My mother* [eventually] *bought one. It was very expensive at that time. There was no use for it other than as a hobby.* Gniibe was fourteen when he first used a digital computer. *It was just difficult for normal people.* Popular computer magazines printed code lists that Gniibe typed up in order to play games. Around the same time, he learned about a new hobby in a popular computer magazine. *There was a special issue about the Rubik's Cube. At the time, I visited the local university for a festival. The ham club had a computer there as well as a Rubik's Cube and a solving memo. This is why I visited the club.* It was this occasion that enticed his brother to acquire a personal computer. Gniibe does not remember losing sleep because of an obsession with a computing issue, but he has vivid recollections of having trouble falling asleep because of the Cube.

Gniibe attended a public school in Maebashi and later informed his parents of his decision to attend Gunma Technical College for high school. The decision to major in electronic engineering came naturally as he already self-identified as a ham radio guy by the time he transferred to the college. *Years ago we had a regulation in Japan prohibiting commercialization of radio. By building radios ourselves, we did not have to pay government tax. So radio kits became very popular in Akihabara. We had radio kits and ham radio. You know, for ham radio you need a license. For just the radio, you do not need the license. In the late '70s, late electronics age, we had many interesting projects with transistors and integrated circuits. We could build many things with them. I had such a hobby. I was thirteen years old.* Another important reason for his decision to enter the engineering department was to be away from his brother. *Our age difference is only one and a half years. I do not have memories of us playing together, but I do have many memories of us fighting each other.*

Gniibe was a dedicated student who had a habit of questioning his teachers. One instance was particularly memorable: a teacher refused to give him a technical explanation, telling him that what was important was not the mechanism but the fact that someone had invented the transistor, an explanation eleven-year-old Gniibe felt to be far from satisfactory. *I remember that I asked him about the mechanism: how do transistors amplify sound? That was my question. But I was disappointed. I remember the answer. He said it was not so important how they amplify sound. It was more important that we invented the transistor, so we can amplify sound. It was a human being who invented it!*

By the time Gniibe turned twelve, he was appointed to be the chairman of the student board. He attributes this fact to his vast milk-cap collection. His experiences collecting and trading milk caps now shed light on his adult experiences with common circuits. Through his early engagement with exchange practices, he came to perceive a divide between the parallel worlds of the internet and what he describes as "real life": *When I was eleven or twelve, there was a boom of milk-cap collection in Maebashi. I was the king [laughs] because I invented many ways to collect milk caps. At that time, I had at least 10,000 milk-caps. I visited milk production plants. The power of a school student is very limited, but I went to many places around here. I had relatives who had jobs in different prefectures. So I asked them to send me milk caps. Usually, you have the address of milk vendors on the top of the cap. So I wrote them for other kinds of caps. I was a kind of millionaire of milk caps! [laughs] You know that usually kids in elementary school only get a cap a day. I got hundreds a day. But this is just half the story. As I mentioned before, I had an allergy and my doctor used to be a milk-cap collector twenty years ago or so. I learned that the boom was a kind of cycle. So, I got a huge collection of ancient milk caps from him! You know, ancient milk caps from Fukushima were more valuable. I controlled the market. It [was] a parallel economy. My principle was that milk caps are very important; we do not exchange them for other things. Only milk cap for milk cap. I could have been more rogue and exchanged my milk-cap power for other kinds of goods, but refused because I loved milk caps very much. At that time, I wanted to preserve this parallel economy intact.*

As a trader, what fascinated him the most in regard to computing was a deep sense of power.[5] *I felt that I could do anything in cyberspace; I have absolute power in cyberspace, but my friends have absolute power too. You can create your own world, even though it is a very limited world in comparison to the real world. When I say, "Do that," the computer just does it. Even so, it is very difficult to tell it how to do "that." It is very difficult to use this power. The computer does what we write, not what we intend. It is very difficult to serialize or describe our intention in machine code. It has been fantastic to translate [intentions]. It has been challenging as well, but I still enjoy it.* The nexus between control and pleasure was the outcome, not the point of departure of a long process of self-training. *We need to control ourselves before trying to control something.* Self-training involved a discipline of self-control as a condition for becoming a moral person: *When I use a computer, I feel its power. When I write code, I feel its power, too. Perhaps that is because I am not a native computer user—I*

learned!—or a native programmer—I learned how to write code consciously with
determination! For me, the computer is another existence; it is not my friend.

As a university student, Gniibe learned about low-level computing by
assembling a machine from scratch based on a Motorola 68000 kit he acquired in Akihabara, Tokyo's famed electronics district. He also worked
on the GNU Emacs editor and collaborated on the internationalization
for Japanese language input with other Japanese developers. Computers were incredibly expensive at the time, so people shared software and
computing power in time-sharing and multiuser operating systems. *Back*
then, we did not have free operating systems. Software was distributed in source
form, because Unix was not that magical: for each installation we had to adjust
the code. Every student did that. In 1990, Gniibe found a bug in GNU Emacs
and posted a report to gnu.emacs.bugs newsgroup. To his surprise, a Japanese developer responded to the bug report and the response was positive,
but not very exciting: "In the upcoming version of Emacs, the bug will
be fixed." One year after this first contribution to a Free Software project,
Gniibe became the network administrator and Usenet postmaster for the
Mitsubishi Research Institute (MRI). *In January 1994, MRI connected to the*
Internet through the first internet service provider. At that time, all machines
were GNU/Linux. . . . For me, comp.os.linux was so important. We had a users'
guide of JUNET [Japan University/Unix Network], *which explained the cul-*
ture of cyberspace and its "acceptable usage guide" but many new users didn't
care and caused trouble. I got complaints from another postmaster and I thought
he was right. We had trouble with international e-mail exchange. Although I was
the postmaster—that is, the person in charge of the MRI network by the defini-
tion of JUNET—I did not have power to educate MRI users to follow the rules
of JUNET. I was just a research associate. Many people just asked me for "ser-
vice" and "technology" and they didn't care about culture and rules. Perhaps, this
would be the reason why I became a fighter to protect cyberspace culture.

JUNET proved to be an important platform for participation in the
Usenet, an early community-driven virtual space, self-organized around
antagonistic sentiments in respect to bureaucratic and top-down hierarchies of information management. Anthropologist Bryan Pfaffenberger
described the original goal of the service as disseminating information
on how to use Unix, but later it became a space populated by users who
saw it as a new frontier for "freedom of expression" about taboo topics.[6]
JUNET was Japan's gateway to the early transnational fora of technical

discussion and development in which Gniibe would come to participate actively. *In the '90s with UUCP* [Unix-to-Unix Copy] *network protocol we had an exciting world. JUNET required consciousness to be a network member. These days no mobile phone carrier requires this sort of thing. Users are just consumers from their point of view. When joining the computer network of JUNET, we had a guidebook for computer network; newcomers had to read the manual or else the members would say, "Read The Fucking Manual (RTFM)!" We had to follow certain rules. Most people had to care about the network itself; people had to care about cooperation. People had to understand how engineers were working behind the network. These days no consumer cares about the engineer behind the net. People think it is a consumer right to demand.*

From 1993 to 1999, Gniibe worked as a Linux kernel developer, maintaining drivers and porting the operating system kernel to work with a low-cost Japanese chip, the SuperH processor for embedded applications manufactured by Hitachi. By the mid-2000s Gniibe became disillusioned with the Linux development community because of its position in respect to more general questions regarding computing and society. He decided to quit Linux development because, in his words, "the project was not based on philosophy; it was technology-centric." Another reason for quitting was the decision of the project leader, Linus Torvalds, to adopt a proprietary software, BitKeeper, as a source code management solution for the development of Linux.[7] *When I joined the Kendo club in junior high school, my expectation was a kind of philosophy more than technique. When I joined the internet, we were taught about its philosophy. For me, it is the same with Free Software and the GNU Project. I realized that I was totally wrong about Linux. I expected something similarly philosophical when I joined. Linus's interpretation of GPLv2 is peculiar. He only focuses on his buddy developers and, in my opinion, ignores the power of software developers over their users. I think that relevant technology comes with relevant philosophy.*

When Gniibe speaks of his experiences with different Free Software communities, he does so in relation to his experience as a milk-cap trader. *A similar thing happens around the Linux kernel. It is kind of a parallel world. At least at the beginning of the Linux phenomenon, no one thought about the real business of kernel development—at least not Linus, who started the Linux project as a hobby when he was a student. We exchanged our code and we believed it was quite valuable, but for me, it was sort of a milk-cap exchange. If I was Linus, I would have as a principle not to exchange the value of the kernel in real business. But Linus had a different idea. For me, milk-cap exchange and source code ex-*

change look similar. You know, it is only boys who exchange milk caps and source code. Here Gniibe makes explicit what we could call, after anthropologist Marilyn Strathern,[8] the "gender of the Free Software gift"—in other words, the gendered experience of gift exchange that involves engaging in argumentative quarrel on the register of masculine performance of technical prowess and public display of technical knowledge.

The Linux community was a source of unpleasant memories for Gniibe and other developers, mostly women-identified and Free Software novices who felt threatened by the adversarial environment created by opinionated experts. *My worst experience was during the SuperH porting project. We had a conflict involving the Sega Dreamcast console support. Technical information was not easily available. Basically, only engineers who signed the NDA could get the information from the manufacturer. We got mysterious contributions from anonymous persons and had discussions about how to include them. I was concerned with the source of the contribution and [their] responsibility. What if we are sued by Sega? We had a flamewar. Some people even talked about the quality of the code. Each person had a different viewpoint. It was a hard experience. We did not get anywhere. The result was just bad feelings toward each other. Sometimes it is easy to demand but tough to do things. Around the Linux kernel [project], some people think that a rogue, aggressive attitude is cool. I do not think so.*

The question of computer security—which occupied a significant portion of Gniibe's career—involved, in his words, "caring for human beings, not only caring for computers." *What are the most common errors programmers make? For those things we have to analyze human beings. This area is a bit difficult for me.* In 1999, Gniibe became chief of computer security of the Computer Emergency Response Team Coordination Center (JPCERT/CC), the Japanese counterpart to the CERT that was originally despised by the hacker underground in the United States.[9] From his vantage point in the high caste of international Free Software developers, Gniibe had the opportunity to contemplate various forms of hacking. The practice itself from his point of view could be defined by the criteria of "inventive usage" of available technical means, not trivial, reproduced, and mimicked technical practices. *I spent less than a year with the JPCERT/CC team and based on that experience, I could not find any hacker among crackers. They just mimic! Basically, they are script kiddies. We should never use the word "hacker" to describe people who just mimic. A hacker does creative things. People who just mimic are not hackers. It is important to just follow and copy at the beginning, however.*

It is an important activity to mimic the things which were already done in order to learn. The evaluation whether an activity qualifies as "hacking" should not be done in terms of influence or effect, as it is common to identify the power of a single person, usually a young person, in bringing down the service or the network, rendering momentarily powerless, say, a huge corporation or a governmental body. The contemporary enactment of the mythical battle between the young hacker David and the powerful corporate Goliath does not meet the criterion of a hacker practice as such a battle does not involve poiesis.

Even though Gniibe identifies with Zen Buddhism and Shinto, he does not readily recognize a significant influence of his religious ideas in his day-to-day technical practices. *I think I am a typical Japanese. I go to the shrine and I go to the temple and I do not distinguish the two too much. I do not identify or distinguish my practice or ideas if they come from Buddhism or Shinto. I carry other values, but it is not because of Buddhism.* Elements of his moral guidance, as he elaborates on them, consciously come from the Buddhist spiritual discipline found in the Soto school. When it comes to his work as an engineer, Gniibe applies the approach of *not* using his technical powers. He does not impose or try to influence people's decisions in adopting Free Software technologies. *For me, expecting or asking others to have the same state of mind as myself is a sort of violation of human dignity. On the other hand, maintaining this internal condition for myself is a matter of coherency and integrity.* His work for Free Software involves what he calls an "indirect way" of engaging in a social movement: it is by building technologies that do not trap users that people get invited to participate. It contradicts a notion of relationship in which programmers are vendors and users are consumers. *Our power is programming, so we do our best with our tools. Our purpose is the social movement. We do not force users to use the GNU operating system. It is not an exercise of our power. We do not enforce—it is up to user to select the GNU operating system or not. It is OK for the user to just use and not join the movement. We do not exercise our power over users. At least, I don't.*

Restraint in the use of power has other ramifications in Gniibe's life. It extends to his immediate family circle and the way he deals with the education and entertainment of his children. During a visit to his home in Maebashi, I noticed how much he disliked Disney-themed toys and cartoons as entertainment for his children. He affirmed he would not prohibit them, but he would not encourage or invest in them. He reasoned that Disney is one of the key corporations lobbying in the United States for the advancement of copyright extensions, which are also being forced upon other countries via international trade agreements. I also noticed that his partner's computer was running Windows, so I asked why. Gniibe launched into an explanation of the national fame of women from Gunma as strong matriarchs. There was little he could do about her choice of operating system.

In 2002, Free Software Initiative Japan (FSIJ) was founded by a group of computer professionals accompanied by a business professor in Tokyo. From the beginning, Gniibe's involvement with FSIJ for the promotion of Free Software in East Asia entailed a constant struggle, but it was also an opportunity for ethical self-reflection. When it comes to the dispute between Free Software and Open Source over tactics for attracting more participants to software projects, Gniibe sides with Richard Stallman. *Many people abuse the term. The Open Source Software definition itself is good. In the name of Open Source, many people did bad things, I think, hiding problems, hiding conflicts, just talking about one aspect, the good aspect of technology. These people do not talk about the practice and cause trouble. I do not think it is fair. At first, we should be explaining our practice to some extent. Most people explain good points and do not explain bad points. In Japan people use the OSS abbreviation. The usage of OSS is a kind of a nightmare for me because these people do not care at all about its full definition.*

Japanese companies were early adopters of Free Software alongside US companies that were created around software distributions and customer support in the mid '90s. The appropriation of the term *Open Source* in Japan has a different trajectory, however. It was adopted with a similar meaning, written in *katakana*, based on the collaboration of local developers with US-based companies and Open Source Software initiatives. It also had a different trajectory within the governmental sphere. *You may know that in the past, 2002–03, when we got the funding from METI, it was for an Open Software Project, because one of the top officers did not like the Open Source and the Free Software ideology. The next year, the president of IPA understood the Open Source movement. After that, another top officer joined the project and changed the name again. It is the same as Open Source Software, but they have a different name* [Open Software, written in katakana]. *The top officer did not like having an external organization to control government planning. He did not like what was not invented here in Japan.* During his time as a government employee, Gniibe managed to achieve a goal he had set for himself. *My challenge as a government employee was to change the reputa-*

tion of the Free Software movement. And I had some success. These days it is not considered wrong to participate in Free Software development. Today I can say I am doing Free Software development proudly! But twenty years ago, it was a bad thing to say. Gniibe held his government position as a research associate for a decade. He confided that he commanded a good degree of respect and freedom within the government given his technical knowledge, which allowed him to dedicate his time to Free Software.

As a long-time contributor to the electronic commons, Gniibe's study of the legal code behind Free Software licenses came late in his trajectory. It was only a decade after he submitted the first patch to a project that he came to familiarize himself with the intricate details of software licensing. In 2006, the Free Software Foundation in United States organized seminars in various countries (with vocal Free Software activists and strong Free and Open Source Software communities), such as Brazil, India, United States, Spain, and Japan) to discuss the update of the General Public License to its third version. Gniibe not only organized and managed to get funding for the meeting in Akihabara, Tokyo, but also participated in all the seminars and in the training in New York with the lawyer Eben Moglen, director of the Software Freedom Law Center.

When asked about his relationship to his work in terms of ownership, he describes his position as unusual in comparison to other developers. Certain forms of naming, Gniibe suggests, but not ownership are considered acceptable for cases that are similar to those in mathematics where one theorem carries the name of a researcher. Naming and claiming ownership of circuits is something that Gniibe does not recognize as legitimate. Software stands in a different category; it is not completely authored, but it is not as anonymous as a circuit design either. *For schematics and electronics, we do not put attribution for inventors. I think it's just part of the culture. But for me it's OK to put some attribution as a way to acknowledging an author, though I do not think it is ownership, especially for schematics or printed circuit design. On the other hand, I feel a sort of ownership; like a story or literature or technical paper, we have a sort of ownership. Software is something between schematics and manga, literature, and academic paper. In some ways, I would like to demand that these are my words, this is code that I wrote. But it is OK for me to have people modify my code. In an article, it is my opinion. It is OK to change the text to write your own opinion using my text, but then it is no longer my opinion.* Learning about the process of revising and debating changes to the GNU General Public License (GPL) became an educational project on the challenges the Free Software community has faced over the course of three decades. *In the process of drafting the GPL version 3, I learned how it works within Japanese law. Actually, inside the GPL we can find some of the history of the Free Software movement. We have many methods for preventing binary-only distributions, a practice of proprietary software and some terms against Digital Rights Management, some terms to work around Digital Millennium Copyright Act, some terms against software patents, some terms against Tivoization,[10] and so on. We do not need to understand the legal detail, but we can learn much about the Free Software movement inside the license.*

Hacking conceived as a novel, inventive practice takes not only technical ingenuity but also an element of self-discipline and adjustment of one's actions in respect to others. Gniibe's "libre hardware" project illustrates his conception of hacking as embodied in the digital tools of his own making. Gniibe started working with the GnuPG (Gnu Privacy Guard) project after realizing that the lead developer needed serious help, and he soon became well versed in cryptography. Gniibe struggled personally with what he identified as a common misperception in the security industry: the perception that the user is generally a malicious agent. He therefore started to work on the implementation of cryptographic functions on a low-cost chip that he found in popular kits for engineering students. Using this particular chip, he designed a substitute for the industry standard Smartcard, a USB token to be used with GnuPG for data encryption, cryptographic signatures, and authentication. The device was a hack in itself from various angles: first, by transforming a device for educational purposes and using it for an application in the context of privacy and data security; second, by rending Smartcards "transparent" (that is, a device that can be inspected by users not to probe secrets but for educational and auditing purposes); and finally, as a combination of Open Hardware and Free Software, articulating two domains of common development that have their own histories, methods, and established communities. The whole design of the crypto token was made by Gniibe and sent to Seeed Studio, as we saw in chapter 3, completing a cycle of design, implementation, and concretization. Most importantly, the token illustrated what is at stake for Gniibe when it comes to hacking: an inventive, creative practice that serves as a vector to provide "transparency for the users."

Over three decades and across several projects of key importance for the computer common, the golden rule, as articulated in the GNU Manifesto,[11] has been contemplated by Gniibe in the negative form expressed

in the Confucian tradition: do not impose on others what you do not wish
upon yourself. This difference in moral articulation accounts for a crucial
distinction in which Free Software and Open Hardware development are
articulated in Gniibe's experience. One of the key moments of ethical reflection on computing issues happened after his first encounter with the
Japanese version of the GNU Manifesto, which he came across in the ap-
pendix of the JUNET manual around the late 1980s. The manifesto is a
key document in the history of computing for articulating the question of
software sharing in terms of a moral injunction. For some technologists, it
is a foundational document; for others, the usefulness and quality of Free
Software programs were the most enticing factors turning them to collaborative software development. *Reading the GNU Manifesto, I thought about
the dilemma and I wanted to support Richard* [Stallman's] *idea. I thought it was
the right thing. But the real world is different; you see that* [there is a common
practice in the industry] *of not distributing the source code. Then, how should
I proceed? That was the question.* In the context of the Japanese computer
industry, Gniibe considers himself somewhat unique in his commitment
for thinking deeply about the ethical implications of his knowledge. He
identifies an inertia among his fellow Japanese developers in their avoidance of accepting what he calls the "structure of the dilemma" he found in
the Manifesto. *How should I not apply my power when I write code? That is my
question.*

CONCLUSION

Are Hackerspaces Prefiguring Technopolitical Alternatives?

What would computing look like if it was not overpowered by the utilitarian logic of the CEO, the IT manager, the IP lawyer, and the financier? What could computing become if it were reinvented through convivial circuits in the spirit of the Maori proverb (transcribed in the conclusion of Mauss's *Essai*) that recommends "giving as much as taking" for "everything to go well"?[1]

Before entering college, I held an eight-to-six job as a computer technician for a small company where I had my first experience with the transnational anti-common. My job was mostly uninteresting for those who could not care less about computers. I was responsible for fixing computers and supporting a local network, making sure everything was running as it should. One morning, the company was brought to a complete and unexpected halt with the impromptu visit of the federal police accompanied by a computer science professor serving as an expert auditor. The raid was led on behalf of the Brazilian Software Association—with long-distance puppeteering by the Business Software Alliance in Washington, DC—to audit computers and identify illegal copies of proprietary software. Under the injunction of a search warrant, I was asked to accompany the auditor machine by machine and take screenshots of serial numbers for every piece of software installed. When we approached the computer room where I worked, I could not help letting them know how eager I was to have them inspect my machine. At the time I had a literal Frankenstein on my desk: an ordinary i586 but with an expensive disk controller (Adaptec 19160) and a blazing fast SCSI disk. I kept my machine without the hood on, so that I could substitute its internals with minimal effort. As my computer ran only Free Software, I proceeded to tell the auditor half-jokingly: "I bet you'll have fun inspecting this one!" To my surprise, he decided not to inspect it at all as he had also not bothered to verify the servers, which were also running only Free Software. This event was my first practical lesson on transnationality, computing, and power.

Fast-forward two decades. We now find ourselves in a condition where the project for the "digital commons" has definitely "succeeded," but mostly for infrastructuring[2] corporate actors. "Big computing" has finally realized many of our anticipated nightmares of mass surveillance and radical monopoly, all powered, curiously, with open technologies in every domain that internet companies do not deem amendable to advertisement or rent collection. Ubiquitous and data-intensive computing started to enable unprecedented levels of extraction and, subsequently, displacement, which social scientists had primarily identified with early cycles of capital accumulation. After decades of relentless struggle to create alternatives to the intellectual property regime, open technology development became an unavoidable corporate reality, serving as infrastructure to bootstrap companies to the detriment of independent communities of technologists. As key symbols, "sharing" and "collaboration" entered corporate discourses quickly. Open technologies became ubiquitous, running on most servers and personal and portable computers, becoming invisible in the process (with the subsequent invisibilization of those who contributed the work). They became strategic and convenient for "agile software development," where the common is often mistaken for de facto industry standards and corporate interests in interoperability. Meanwhile in community projects, contributors were burning out from overwork as well as from recurrent discrimination and violence (mostly symbolic but also physical) brought to public attention in mailing lists and blog posts in the form of collectively signed manifestos. All the while, internet companies continued to benefit from overwhelmingly rich repositories of code, "forgetting" to give back, thus breaking the spirit of the unpayable gifts of software and hardware that have been historically given. This was made public and notorious by Eric Schmidt in his advice for aspiring "unicorns" in startup businesses: "Open is not a moral argument," he maintains. "Defaulting to open is usually the best way to drive innovation and lower costs in an ecosystem, so view it as another strategic tactic at your disposal: Will going open help you achieve scale and profitability?"[3]

Yet, this is just one aspect, albeit central and overpowering, of our technopolitical condition, the result of heavy ideological investments of neoliberalism as a globalized-yet-provincial capitalist cosmology. As this book has ethnographically delineated, there is an alternative side to the globalization of computing expertise, but also and most importantly, there is another way of learning to learn about computing through common circuits. In the preceding chapters, we identified something essential that brought together activists, technologists, and scholars who worked on the question of the common: the affirmation that the common is, in fact, everywhere to be found. We encounter its reaffirmation as a political prefiguration in ongoing efforts to create anti-surveillance alternatives to the regime of large-scale data extraction and accumulation in many hacker projects. We find its realization in forms of self-governance in hackerspaces with every attempt to interrupt the march of rampant privatization and individualization. We find the common in every concrete act of solidarity to form networks of mutual aid that places "people first, above property and profit," as an anonymous Noisebridge stencil proclaimed. And we brought all this together ethnographically with the concept of *common circuits* to describe the technical, social, economic, and spiritual entanglements of technologists with open technologies. We learned that it is a matter of identifying and fostering its concrete spatial instantiations: sharing community spaces and hardware in Shenzhen, San Francisco, and Tokyo with parallels in the practices of sharing heirloom seeds in India and Brazil, building food and housing cooperatives in Spain and Portugal, and creating local exchange trading systems (LETS) with social currencies in Canada and Argentina. Experiences of commoning are too numerous and too widespread to recount, but they are also too often made invisible for not fitting the prevalent narrative of infinite growth and uninterrupted scale. The academic critique of "neoliberalism" (comfortably armchaired as it has become) has prevented many of us from seeing and realizing concrete alternatives. And this inability to perceive and promote alternatives has led the defense of autonomy to be mistaken, in the end, for rugged individualism, mutual aid to be mislabeled as charity, and self-organized community-making initiatives to be disparaged as "infantile leftist idealism" that "does not scale."

In this book, we have engaged this debate concerning technopolitical alternatives through an ethnography of common circuits populated by technologists who identify themselves and are identified by their peers as "hackers." Our guiding question concerned the alternatives they embody in contrast to the experience of computing in educational and corporate sectors. We identified in the commoning practices of gift-giving interruptions in the logic of neoliberal economization that turns everything into property and everyone into restless entrepreneurs of the self (and others).[4] I borrowed the language of interruptions from the work of anthropologists Lisa Rofel, Aihwa Ong, and Anna Tsing on the cultural manifestations of neoliberalism within and outside the Euro-American axis. In the work of these anthropologists, we find active questioning of transnational capitalism as an incomplete and profoundly unjust operating system with its interrupts, routines, logistic pipelines, and system calls. Our transnational condition is now better understood—with anthropological studies on "globalization from below"[5]—as a heterogeneous systems network, running on different platforms with their own legacies, interrupt-handling functions, and degrees of autonomy and obligation in exchange circuits. Hackerspaces sometimes enable interruptions in the corporate circuitry of computing, but they also are instrumentalized to foster neoliberal projects. They are far from immune from the dominant technoscientific circuits, as they draw from them many of their technical objects, projects, and constituents.

The three registers of *spatialization*, *personification*, and *politicization* discussed in previous chapters were devised to put into evidence these unlikely interruptions: (1) technologists as hackers animating exchange circuits among strangers on the register of the gift; (2) self-cultivation for participation in the moral circuits of open technology made possible through commoning; and (3) mutual aid through collaborative arrangements challenging the persuasive logic of association around entrepreneurial forms. We saw that conviviality can be fostered to interrupt the calculating logic that shapes most computing projects, privileging instead collaborative learning relations that are often not framed as such.

In the ethnography I also signaled the concrete difficulties of bringing about technopolitical alternatives as interruptions because alternative sociotechnical experiments were entangled with the production of inequality within and beyond the field of computing. The will to create community, we found, was frequently obliterated as technologists reproduced hierarchical relations, often in paradoxically exclusionary modes of commoning among "equals" that hardly extended beyond small groups of experts. In the global circuitries of open technology, the common has always been characterized by aspirational practices and marked by intersectional frictions. It has represented historically and geopolitically much more of a *commons in the making* with substantial barriers to entry for those who are not cultivated in computing. And yet, despite all the challenges, open technology collectives also manage to swim upstream and create the *common in practice*.

We saw in previous chapters that the pendulum of moral valuation has oscillated historically between the "hacker" as a shadowy underground figure who is behind computer-based crimes and a virtuous technologist who mobilizes technical knowledge to improve digital systems through community-making. In the past decade, the pendulum has made another full swing toward the positive side of the force with the formation of a network of hackerspaces, a global but partial circuit of heterotopic spaces for learning through digital experimentation. Against the backdrop of informational capital, we witnessed an "impossible" practice emerge across heterotopic sites to create the common out of shared spaces, tools, techniques, and technical objects. The novelty here resided in extending the electronic commons to physical spaces and tools. In San Francisco, Noisebridge mobilized an undisciplined collective against its surrounding corporate computing culture by experimenting with "radical openness"—a heterotopia set against the technopolitics of Silicon Valley where "sharing" implicates what political scientist Wendy Brown identified with "un- or under-employed subjects who cannibalize to capitalize their homes, bodies, cars, bikes, spare time and other 'assets' for survival."[6] In the overclocked circuits of Shenzhen, Chaihuo constituted a space of ephemeral suspension of market pressures to create pathways for an economy of Open Hardware–based gifts among local and foreign engineers, but with serious limitations for other kinds of "makers" on the factory floor. At Tokyo Hacker Space, conviviality around open technologies provided the necessary catalyst for an unexpected collective to self-organize a serious response to the nuclear disaster. In all these instances, concrete community spaces served as exchange hubs to prefigure technopolitical futures: they instantiated a common circuit with hackers and open technologies not only *alongside* but also *against* institutional spaces of computing in universities, industries, and governments.

If *spatialization* is a key register for understanding the contemporary politicization of computing expertise, the other, I suggested, is *personification*: how technologists cultivate themselves as hackers in the context of the open technology circuits. Although the hacker identification and stigma have been primarily built upon the Euro-American "globalized provincialism" (to use the felicitous expression coined by sociologist Renato Ortiz), the trajectories of Gniibe, NalaGinrut, and Mitch Altman were selected for their influential rearticulation of different moral idioms, under contrastive conditions of possibility, of what it means to be a "hacker." Their distinctive definitions of hacking did not create impediments to collaboration and participation in the common circuits, but they exposed quite distinctive and unequal conditions for self-cultivation. The demand for self-training in hacking found expression in local religious and political traditions with important implications for the three personal trajectories we examined, rendering one's engagement in computing collectives meaningful, justifiable, desirable, and honorable.

When NalaGinrut articulated what matters most to him in terms of community-making, he was not unreflexively echoing something he had heard online from old-school hackers in the Lisp community. He was reacting to the rampant individualization in Shenzhen, an environment with the world's fastest electronics rat race ("a race to the bottom") for the young engineers of his generation. To dedicate oneself to the work of independent community organizing in this challenging context is to create an interruption, albeit temporary and fragile, in the imported logic of "move fast and break things" that startups all around him embrace. As we saw, he is in no way anti-commercial. Quite the opposite—many of his projects are also intended for the local and international markets. But for him, the centrality of community as vocation is still very much part of his technical cultivation at the crossroads of neo-Calvinism, neoliberalism, and neo-Confucianism. It is symptomatic of our times that the substantial work that Nala and his fellow hackers from SZDIY have contributed to create community around open technologies in Shenzhen goes largely unnoticed. They cannot be found in the news, magazine articles, and governmental reports about making as a "road to entrepreneurial success" in mainland China.

When Mitch Altman spoke of love, he was not being spoken through a PR campaign of a major Silicon Valley company as most technology critics are quick to assume. His positioning needs to be contextualized with respect to his trajectory of rejection of the overwhelming presence of the military in science and technology, his disappointment in his experiences as a startup founder in Silicon Valley, and his vision of community building as, ultimately, internationalist and anti-capitalist. Mitch has been the informal ambassador for the informal international network of hackerspaces but has seen the project co-opted by Silicon Valley and the Chinese government under the rubric of "makerspaces" as entrepreneurial innovation spaces. He sought tirelessly in his career for a technology project that could not be incorporated by the military. Eventually, he found one: an open countercultural device that is meant to turn off the propaganda machine.

And, lastly, for Gniibe, hacking assumed unfamiliar contours as an ethical practice meant to prevent him from exercising his power over others. In his inversion of the golden rule, he signaled a form of technical commitment that is, above all, spiritual. It has to do with his understanding of his place in the world as a person of technical ability in relation to people with different abilities. In his trajectory, openness figured as a means for building and inhabiting moral circuits with different response-abilities. Gniibe is one of the numerous invisible engineers who built not only the basic operating system code that billions of computers depend on but also the cryptographic technologies to secure the communications of whistleblowers, such as Edward Snowden, in their acts of civil disobedience. The fact that his work does not figure in books, magazine articles, and news reports about "crypto" only demonstrates the urgency of attending to minor circuits of computing.

From the personal and collective trajectories we traced in this book, we may now be able to identify the common circuitry that led to the prefiguration of a new technopolitics. In terms of the broader consequences of this argument, the most pressing concerns the anthropology of the political that is not yet identified as such, the *technopolitical otherwise*. Beyond classical analyses of cohesion and conflict in political anthropology, the hackerspaces network stands as an "impossible object"—that is, an assem-

blage with varying degrees of openness across insurmountable socioeconomic and geopolitical differences and distances.[7] It could not be different:
hackerspaces are all part of distinct sociotechnical landscapes with their
own political histories, even though they display "family resemblances"
as suggested by Isabelle Berrebi-Hoffmann, Marie-Christine Bureau, and
Michel Lallement in their reflection on the importance of these experiments for reinventing work relations.[8]

Taken as an expressive set of ongoing sociotechnical experiments, we
can identify similarities in hackerspaces that overlap and crisscross but
also rapidly disappear as socioeconomic distances surface. Rather than a
comparative exercise, we saw that it is more generative to focus on network forms, efforts, and effects to examine technopolitical phenomena
in relational ways. The participatory devices we described in previous
chapters were designed for "openness" and "collaboration" as ethical
stepping-stones to sustain a minimum degree of coordination for prefigurative work. Making hackerspaces "open" for poiesis, as they aspire
to be through their "design patterns," has been much more difficult than
it is publicly portrayed. Performances of technical expertise are key for
participation, but they also constitute major points of contention, if not
outright exclusion. One needs to engage in public performances of technical knowledge to be considered "worthy" for a hacker project, which, of
course, leaves a lot of aspiring members of minoritized backgrounds out of
the picture. This tension is constitutive of the exercise of a non-hegemonic
masculinity among computer geeks, consisting of recurrent agonistic or
amicable arguments about technical matters. I have found this practice to
be very common in community spaces, as ethnographers in the context
of neighboring technical communities have also reported.[9] The question
of the common always elicits further questions: Common for "whom"?
Common of what and under what historical circumstances? As an SZDIY
member once told me at the limit of his experiences, a limit expressed in
many corners of hackerdom, "I know there are women hackers out there,
but I have never met one."[10]

In locating alternative manifestations of computing expertise, we
have seen how hackerspaces were co-constituted at cross-cultural contact
points to rearrange, challenge, and transform technical practices, more
broadly, through the key symbol of "hacking." In their distinctive moral
circuitries, hackerspaces delineated the negative contours of credentialed
computing expertise. Through sociotechnical ties, they re-created a form

of alterglobalized solidarity among computer experts in the most unlikely conditions of fast-expanding individualization, competition, and
anti-commoning at transnational scales. In the end, the anthropological
relevance of hackerspaces resides in their support for renewed technopolitical arrangements: creating concrete alternatives is a matter of realizing
network potentials through practical experiments at these emergent and
interconnected convivial spaces. Hacking *ties* technologists to the circuits
of communal experimentation, but it also serves for *hacking* ties with organizations that are perceived to be detrimental to common circuits. It is
within this productive ambivalence that hacking sustains itself as an undisciplined but convivial technopolitical practice, for every common project gestates many potential and actual others as gifts that keep on giving.

Notes

Introduction

1. The *commons* is most frequently identified as a politico-economic framework that stands as an alternative form of governance to markets and states, offering self-organized institutional arrangements for managing shared goods (Ostrom 1990). The *common* in its singular form, however, refers to the recent debate on *commoning* as a political practice that is not limited to the affordances of "natural resource systems" that Vincent Ostrom and Elinor Ostrom (1977) identify as "common-pool resources." The theoretical move away from the definition of the *commons* as resource is identified as particularly pressing for recuperating the meaning of commoning practices that create de facto politico-economic, technical, and ecological alternatives. This move is also defended as politically urgent in foregrounding the move in the historical understanding of dispossession related to land enclosures and other natural resources toward intangible goods (with intellectual property enclosures and, more recently, large-scale data extraction and rent collection in digital platforms). In this book, I use the term *common* after Pierre Dardot and Christian Laval (2014) and Silvia Federici (2018) to describe technopolitical practices of *commoning*—that is, practices that "institute something in common," politically and materially, for collective self-governance and mutual benefit. I continue to use the term *commons* throughout the book in reference to the public debate on the digital commons—a domain of exchange of "open technical objects" under the rubric of Free and Open Source, Open Hardware, Open Data, and Open Access technologies. By using the two terms I hope to highlight the distinction and the impor-

tance of these two domains of study, bridging areas of scholarship that explore technical and political practices alongside self-governance structures without, it is important to highlight, inscribing the work I present here in the tradition of rational choice theory (Obeng-Odoom 2021).

2. Escobar (2018).

3. Social ties are fundamentally the means through which moral economies of gift giving are sustained. Here I follow the definition of the "gift" (*don*) after Alain Caillé's work to renew the political and economic anthropology of Marcel Mauss. According to Caillé, the economy of the gift is characterized by "any prestation of gifts or services conducted, without guarantee of return, that is meant to create, sustain, or regenerate social ties" (Caillé 2007, p. 124, my translation). Camille Tarot corroborates this line of interpretation when she describes the relevance of Mauss beyond former misinterpretations of the gift as "pure disinterestedness" or "reciprocity." What Mauss offers instead is a "radical reflection of the emergence and nature of social ties, caught between violence, rivalry, reason and obligation" (Tarot 2003, p. 73, my translation).

4. The list of colleagues exploring minor histories of computing is not as long as it should be, but it keeps growing with contributions from historians and media studies scholars, such as Jennifer Light (1999), Nathan Ensmenger (2010), Mar Hicks (2018), Joy Rankin (2018), Charlton McIlwain (2020), and Kevin Driscoll (2022).

5. I use the term "conviviality" (*eutrapelia*) after Ivan Illich's (1973) critical examination of the "radical monopoly" of the technosciences over social life, a situation where alternatives are made difficult if not impossible to conceive. Illich described "convivial tools" as an antidote to technocratic, instrumental reasoning. In his definition, conviviality is the "the opposite of industrial productivity," being first and foremost about "autonomous and creative intercourse among persons, and the intercourse of persons and their environments" (Illich [1973], p. 20). This is key for our analysis here as the philosopher inspired a shift in the 1970s from a discourse on computing as an instrument of bureaucratic governance to its integration in everyday forms of life where technologies are "peopled"—to borrow Michael M. J. Fischer's (2009) expression. In Illich's (1971) argument for "deschooling," we also find a discussion of a spectrum of organizational forms that separates "manipulative" and "convivial" institutions. Here we find his critique of "credentialed expertise" that is more often than not a meager substitute for embodied and situated knowledges. The celebration of conviviality with high levels of skepticism with respect to educational institutions and credentialed forms of expertise has been a constant in hacking. My suggestion is that anthropology can help us identify convivial alternatives in emergent forms of knowledge making and sharing, even if only in fragmentary and prefigurative forms.

6. Gilbert Simondon laid the foundational work for bridging the "big divide" between the technical and the social in his critique of what he called "facile hu-

manism." "Our culture," he writes with reference to one of the centers of philosophical and technoscientific knowledge production in continental Europe, "is one that is based on the systematic error of considering the technical to have an independent existence from human affairs" (Simondon [1958] 2012, p. 9). It is in his philosophical démarche—after Marcel Mauss's and André Leroi-Gourhan's contributions to an *anthropology of techniques*—that we find the conceptual tools to decompose technical objects into processes, resonances, and structures, so as to examine their genealogies and answer the question: How do they transform over time to become concrete, active, autonomous, and functional parts of human activity? This is a generative point of departure for an anthropology within and beyond informatics—one that is best grounded in ethnography for its engagement with naturalist and nonnaturalist epistemologies much more broadly.

7. The concept of *common circuits* was inspired by the relational anthropology of Marcel Mauss, whose contribution has been a lasting one for identifying the imbrication of social, spiritual, and material patterns in the circuitries of exchange. In examining the dynamics of gift economies, Mauss identified what sustains their generative character in fostering social ties: they concern magical and technical practices, aesthetic sensibilities, political mediations, moral orientations, and societal obligations forming an intricate circuitry by virtue of circulation of persons and objects. As I quoted in the epigraph, in their "mixing," persons and objects come out of their realms to integrate an emergent process where the divides between the technical, the personal, and the political cannot be empirically sustained. Mauss's key finding of the "triple injunction" of giving-receiving-returning as the elementary dynamic of moral circuits can be found in spaces of conviviality that have not been completely overtaken by market logics (Mauss [1923] 1950; Caillé 2007). As a methodological imperative, one must calibrate one's ethnographic attention to identify moments of gift giving in contemporary circuits since they are quite often mistaken as illogical, residual, retrograde forms of exchange running on obsolete technologies that no longer have a place under the neoliberal order of "big computing." "Retrocomputing," "permacomputing," and "post-collapse computing" are illustrative examples in the everyday life of community spaces that I discuss in this book.

8. Definitions of Free Software and Open Source technologies have framed the debate around two hegemonic Euro-American orientations for digital commoning: the ethics of "software freedom" for the former (under the injunction of the Golden Rule) and the legal and pragmatic defense of the benefits of collaborative software development practices for the latter. In this book I use the rubric of "open technologies" to encompass all the alternatives in the experience of commoning involving software, hardware, data, and other types of information technology that matter to the hackerspace community. *Openness*, in this context, is better understood as a function of expertise (of the technologist qua hacker), the nature of the technical object, and its associated milieu (which

provide conditions for openness to be meaningfully realized). A more suitable term for our purposes here would be *common technologies*.

9. I follow Strathern's theorization of the relations between persons and things in exchange practices: "By objectification I understand the manner in which persons and things are construed as having value, that is, are objects of people's subjective regard or of their creation. Reification and personification are the symbolic mechanisms or techniques by which this is done . . . If these concepts thus refer to anything it is to the forms in which persons appear, and thus with the 'making' (-ification) of persons and things" (Strathern 1988, p. 176).

10. Fischer (2018).

11. Source: US Navy's *Chips Ahoy* magazine (1986).

12. Before the term *computer* was attributed to machines by the mid-1940s, it was used to identify a human occupation. Not only were computers human, but their inhuman activity was mostly carried out by women in key but often unrecognized positions (Fritz 1996; Light 1999; Ensmenger 2010). Active recruiting of women for "human computer" positions led to the creation of a team of early programmers. Several reasons account for what Jennifer Light identifies as the "feminization" of the work of ballistic computing and, with the advent of computer machines, that of the "computer operator." "It is a curious paradox," Light ponders, "that while the War Department urged women into military and civil service and fed the media uplifting stories about women's achievements during the war, its press releases about a critical project like the ENIAC do not mention the women who helped to make the machine run" (Light 1999, p. 473). Both Jennifer Light and W. B. Fritz depicted the artifices used to render women invisible, despite their centrality in training new operators as well as documenting, operating, and maintaining the ENIAC. The practice of rendering women invisible, Light (1999) notes, went as far as cropping them out of the publicity images about the machine.

13. This argument has been supported by historians of computing: "The fact that Hopper wholeheartedly welcomed non-UNIVAC personnel to learn about the A-2 compiler sheds some light on her beliefs concerning intellectual property. Hopper did not view software as a commodity to be patented and sold. She took her cue from the mathematics community" (Beyer 2009, p. 242). William Aspray, Martin Campbell-Kelly, Nathan Ensmenger, and Jeffrey Yost corroborate this line of interpretation when they affirm that "probably no one did more to change the conservative culture of 1950s programmers than Grace Hopper" (Campbell-Kelly et al. 2019, p. 169).

14. Lapsley (2013).

15. See Levy (2010) and Johns (2009) for descriptions of this intersection between phone phreaking and computer hacking at MIT in the 1960s. According to Levy, some of the early hackers were, in fact, quite young, such as Peter Deutsch, who was eleven years old when he first joined the ranks of the informal technical elite.

16. King (1968).

17. It is important to emphasize that this genealogy of "cyber utopianists" became not only mythologized but also deterritorialized, fueling the persuasive magic of "digital innovation" well beyond Silicon Valley. This process has been well explored in the existing literature, so there is no need to recount it here. For a critique of this cultural and political formation, refer to the classic essay "The Californian Ideology" by Richard Barbrook and Andy Cameron (1996), which examines the romantic, individualist, and techno-utopian roots that shaped this hegemonic technopolitical horizon. For a cultural history of this tradition, see Turner (2008). For a geospatial and political critique of its imperial extensions, see McElroy (2024).

18. McIlwain (2020).

19. Ensmenger (2015).

20. The political alienation of many "computer bums" is not only a point of departure for the historical critique leveled at the contemporary celebration of hacking (Ensmenger 2015). It is also a pivotal point for the study of the relation between computing and selfhood pioneered by Sherry Turkle (1984). We owe her for pointing us first to Weizenbaum's work on "computer ethics," but also for providing an empirically-grounded examination of the phenomenology of hacking through narratives of "deep hack" as a mode of complete immersion (with undivided attention) in computing.

21. The question of autonomy is best articulated by Johan Söderberg and Maxigas (2022) through their concept of "functional autonomy" in early networks—including community-driven bulletin board systems (BBS), free-space optical (FSO) communication technologies, and internet relay chat (IRC) service providers. According to these authors, functional autonomy is exercised by self-managed, collaborative control and maintenance of a digital service by the technologists themselves.

22. Illich (1970), p. 93.

23. Driscoll (2022).

24. See Lécuyer (2006) for a comprehensive history of the semiconductor industry in the San Francisco Peninsula. For the urgent (and mostly overlooked) examination of the toxic legacy of this industry, see Pellow and Park (2002) and Gabrys (2013).

25. Pfaffenberger (1988).

26. Thompson (1984), p. 763.

27. Kelty (2008) demonstrated that the computer pioneers of 1970s continued to foster their own circuits through the informal practice of code sharing, animating a rather complex moral economy that involved legal, commercial, technical, and aspirational dimensions. See Salus (1994) and Weber (2004) for additional historical sources on this communal expert practice.

28. Levy (2010), pp. 28–31.

29. Coleman and Golub (2008); Kelty (2010).

30. See Coleman (2012) on the controversy over the distinction between "hackers" and/as "crackers" and Eve (2021) for an examination of underground software distribution networks ("warez") where crackers are regarded as high-status technologists, given the difficulty of their contributed work: "as with other types of computer information security roles, this type of breakage [of software protection] is akin to an incredibly elaborate puzzle that the cracker must solve to succeed" (Eve 2021, p. 103).

31. Important (and controversial) memoirs of this period include *Out of the Inner Circle* (1985) by Bill Landreth; *The Cuckoo's Egg* (1989) by Clifford Stoll; and *Ghost in the Wires* (2011) by Kevin Mitnick. Influential collectives of this period, such as The Realm, Legion of Doom, Masters of Deception, and Cult of the Dead Cow, have been portrayed, respectively, by Suelette Dreyfus in collaboration with Julian Assange (1997); Michelle Slatalla and Joshua Quittner (1994); and Joseph Menn (2019). It took quite some time for hackers to make peace with journalists, but it eventually happened around the time that *hacktivism* started to circulate as a keyword in the late 1990s. This historical trajectory is best narrated by Gabriella Coleman (2016) in ""From Internet Farming to Weapons of the Geek," a brilliant concept she coined by borrowing from resistance studies and adapting it to the political history of hacking.

32. This trajectory was minutely and extensively tracked by Matt Goerzen and Gabriella Coleman (2022) in their research report *Wearing Many Hats* on the spawning of the information security industry from the so-called hacker underground. The authors detail two very important processes in the professionalization of hackers: (1) the rise of the practice of "full disclosure" of security vulnerabilities in important online channels, such as Bugtraq, involving underground hackers and "above-ground" researchers; and (2) public attack and shaming of large corporations, such as Microsoft, for lack of security in their commercial software. Firsthand accounts of the process of criminalization of hacking in the mid-1980s and 1990s can be found in Sterling (1992); Taylor (1999); Thomas (2002); and Goldstein (2009).

33. Farmer and Venema (1993).

34. See Murillo (2020b) for the application of a theory of power in Mauss and Hubert's *Esquisse d'une théorie générale de la magie* to the interpretation of historical encounters between hackers (as magicians) and non-hackers (Mauss 1950).

35. See Scheirer (2024) for the proper contextualization of "fake" materials in underground hacker publications, such as text files and e-zines.

36. McIlwain (2020), p. 96.

37. Marques (2005).

38. Cardoso (2003).

39. "Software Livre" is the term for the sociotechnical experiments that took place from the late 1990s to the present around Free and Open Source projects in Brazil. See Murillo (2010) for the study of its cultural and historical formation.

40. A similar experience has reportedly rung true for several generations of

computer science students who learned about operating systems through Xerox copies of a classic reference in the field, *Lions' Commentary on UNIX 6th Edition, with Source Code* by Professor John Lions (first published in 1976). In my case, the heavily copied book was Operating Systems Design and Implementation by Andrew Tannenbaum, coauthored with Albert Woodhull (and first published in 1987) and known to be one of the most influential references to be challenged by the generation of hackers who gave us the Linux operating system (Moody 2001; Torvalds 2002). The crucial difference here is that many community members in our scene were "unschooled" in computing. For the analysis of the historical relevance of these publications for the Free and Open Source community, see Kelty (2008). For the discussion of schooling and unschooling practices, see Blum (2024).

41. Here I follow the elaboration on "study" as "undercommoning" after Harney and Moten's (2013) work on the possibilities for creating political communion and collaborative knowledge production despite the exclusionary conditions and dynamics of university settings and corporate research laboratories. Following the authors' definition, conditions for studying computing otherwise are gathered in the interstices (through *under*-commoning) of institutional and corporate settings.

42. Zigon (2008).

43. Stallman and Wall (2002), p. 243.

44. For being marginal with respect to the anthropological canon, cyborg anthropologists were deep at work questioning the Euro-American tradition's claims to universality through ethnographic research, demonstrating the situatedness and narrative strategies that have historically been deployed to institute and universalize the figure of the self-centered expert individual working within the well-guarded walls of the citadel of science. That was a tough battle as the field of anthropology was very skeptical of their contributions and reluctant to incorporate their critique. Under the theoretical and methodological orientation of the original Committee for the Anthropology of Science, Technology, and Computing established in the context of the American Anthropological Association of the early 1990s, ethnographies of the technosciences were finally brought to the forefront of anthropological research agenda. See Haraway (1991); Traweek (1988); Downey, Dumit and Williams (1995); and Downey and Dumit (1997) for key interventions that managed to storm the citadel of anthropology (in the United States).

45. After Viveiros de Castro (2015) I take the political commitment of Brazilian anthropology to be grounded in a mode of knowledge production that has to do with the reinvention of the European modernist tradition through theoretical, artistic, and philosophical cannibalism. In his analysis of the "anthropophagic reason," Haroldo de Campos examined the Brazilian modernist metaphor of the cannibal. In his reading, anthropophagy is the "critical devouring of the universal cultural legacy, which is formulated not from the resigned

and submissive perspective of the noble savage" but from that of a cannibal, "which does not involve a submission (a catechesis), but a transculturation or, better, a transvalorization: a critical view of history as negative function (in the sense of Nietzsche), capable of appropriation as well as expropriation, de-hierarchization, deconstruction. Every past that is other for us deserves to be negated. Better yet: it deserves to be eaten, devoured." (Campos 1992, pp. 234–35, my translation). The need for "critical digestion" with de-hierarchization of Euro-American influences, we could say, has been a constant in the reflection upon the place, the identity, and the role of Brazilian anthropologists. The same applies for hackers in the so-called majority world. In addition to its impor-tance for "transvalorizing" philosophies from elsewhere, the cannibal approach has broader political implications for anthropology through its indiscipline: it might help us foster collaborative ties across national traditions, inviting eth-nographers to reestablish collaborative relationships with co-participants that differ from traditional forms of distancing and interpreting through second-hand ethnographic textualization and theorizing.

46. I borrowed this term from Illich's (1973) critique of the effects of the tech-nosciences in everyday life. The basic idea is that every hegemonic scientific and technological state creates a monopoly on the possibilities of engagement with techniques and knowledge, suppressing alternatives by presenting itself as the only way of organizing the search for knowledge, forms of exchange, and learn-ing practices.

47. Fischer (2003), p. 304.

48. Mate (also known as *chimarrão*) is a traditional tea of the American South-ern Cone made from the ground leaves of *Ilex paraguariensis*. Its ritualistic as-pects have been described (briefly) by Lévi-Strauss in *Tristes tropiques* (1955).

49. Toupin (2014).

50. Brown (2016).

51. See Söderberg and Maxigas (2022) for a series of ethnographic cases of hacker projects and collectives that are organized against "digital innovation" as a technopolitical project. Examples range from Open Hardware–based net-working technologies and 3-D printing collectives to autonomous networks of community spaces (such as hacklabs and hackerspaces) and internet network service providers, such as Internet Relay Chat servers.

52. See Ingold (2000) on the topic of "enskillment" and his usage of the eco-logical psychology of Eleanor and James Gibson.

53. "Personhood" is a key but often neglected aspect of technoscientific ex-pertise, but it is central for understanding the distinctive aspects of hacking in contrast to other domains of computing. Important exceptions in the literature can be found in the work of Turkle (1984); Kelty (2008); and Coleman (2012). Hacker "personification" has to do with self-cultivation in a particular socio-technical milieu, as well as with the process through which technical objects rendered as gifts personify the giver in computer expert circuits. Pina-Cabral

(2021) has demonstrated the importance of Mauss for the contemporary study of relations and persons as they are materially and symbolically implicated. "Being as relation" is the *ontogenetic* (more than an ontological) orientation we fundamentally pursue here: from Mauss's classic study of personhood ([1938] 2012) to Simondon's definition of being ([1964] 2017) as "singular common"—a fortuitous expression elaborated by Aspe (2002) in his interpretation of the process of individuation.

54. *Ethical reasoning* is the practice of self-reflection concerning a set of relationships one is confronted with and through which one is, might, or ought to be implicated (Zigon 2008, 2011). It is a form of coping and dealing with contradiction and conflict experienced in everyday life. The late Foucault (1988) of *Technologies of the Self* is one of the key references in this debate, but the most productive framing of the problem comes from the anthropology of morality and ethics, which has helped us to transpose and extrapolate on Foucault's approach to study other historical and political contexts, informed by other elaborations on personhood in contrast with the Greco-Roman and the early Christian traditions (Zigon and Throop 2014; Throop and Mattingly 2018). Ethical reasoning is a fundamental part of the process through which hackers work on themselves as we will see in the chapters that are dedicated to personal trajectories, but it is not sufficient to account for the intimate technical engagement with computing for the formation of hacker personhood. See Turkle (1984), Kelty (2008), and Coleman (2012) for the psychosocial and cultural elaboration on the relationship between hacker selfhood and technical objects.

55. Coleman (2012).

56. Anthropologists have explored manifestations of hacking as cultural re-elaborations on the liberal tradition (Coleman and Golub 2008; Coleman 2012) and pragmatism to create "recursive publics" (Kelty 2008), as well as a generative domain of technical development animated by a shared "moral imagination" of freedom (Leach 2009), technological autonomy, and gift-giving (Zeitlyn 2003; Apgaua 2004; Murillo 2010; Chan 2013). In philosophy and political economy, the most influential contributions have identified in hacker communities the distinctive dynamic of "commons-based peer-production," politico-economic resistance, and social experimentation to create distinct notions of property (Nissembaum 2004; Benkler 2006; Benkler and Nissembaum 2006; Weber 2004; Söderberg 2007).

57. Based on Gestalt psychology, Gregory Bateson's concept of deutero-learning as meta-learning went through important revisions in the course of his work. The definition first appears in 1942 as a response to Margaret Mead's article "The Comparative Study of Culture and the Purposive Cultivation of Democratic Values." In this early elaboration, deutero-learning is articulated as the means for "getting an insight into problem-solving in context." Bateson then provides a working definition: "Let's say that there are two sorts of gradient discernible in all continued learning. The gradient at any point on a simple learning

curve (e.g., a curve of rote learning) we will say chiefly represents the rate of proto-learning. If, however, we inflict a series of similar learning experiments on the same subject, we shall find that in each successive experiment the subject has a somewhat steeper proto-learning gradient, that he learns somewhat more rapidly. This somewhat progressive change in the rate of proto-learning we will call 'deutero-learning'" (Bateson 2000, p. 167). "Learning to learn," Bateson concludes, "is a synonym for the acquisition of habits of thought" (Bateson 2000, p. 166). In his address to a group of psychologists and psychiatrists in 1959, he reformulates his previous observations about deutero-learning in terms of "trito-learning"—that is, as an advanced stage characterized by "learning to learn to receive signals." Deutero-learning is elaborated here in cybernetic terms as "second-order learning" that constitutes the "expectation" that the world (as a system) will be "structured in a certain way" (Bateson 2000, p. 249).

58. It is well established in the anthropological literature that the human body represents a primordial instrument, thanks in large part to the work of Mauss ([1934] 2012) on the "techniques of the body." Yet, under-explored and under-theorized is his definition of *technique* as a "traditional and efficacious act" that is inseparable from other symbolic and material practices of moral, religious, technical, aesthetic, and magical nature. *Tradition*, in other words, is the regular act of transmitting techniques which, in turn, can be considered efficacious due to the very nature of their transmissibility. This debate has been recuperated and updated by Sigaut (2010) and Schlanger (2012). The tacit dimension of this transmission is as important as the conscious, overt, and systematic ones because it speaks to the preobjective conditions for "learning to learn" about computing. This is particularly salient in the context of independent computer collectives: transmission-through-sharing-and-practice is a condition for cultivating one's dispositions (bodily habitus) and also foundational for community building through shared technical objects that, at once, circulate and serve as infrastructure. The feedback mechanism that sustains a "hacker public" through the sharing of technical objects (that become infrastructural for the public itself) has been theorized as a "recursive" property by Kelty (2008).

59. The expression "closed world" was employed by Edwards (1996) in his study of Cold War–era computing and cybernetic imaginaries. I draw from Edwards here to highlight the passage of a particular kind of infrastructural moment in computing toward a phase where "openness" would become a key orientation for neoliberal and cybernetic governance. This understanding of openness, however, is not to be confused with the definitions of the very same term by Gilbert Simondon, Ivan Illich, and André Gorz. It is important to highlight that the opposition Edwards established was between a "green" world of natural, magical, and transcendental forces and the "closed" political and ideological world with its overbearing instrumental reason, allocated to total war with (the possibility of) total annihilation.

60. Nevejan and Badenoch (2014).

61. Riemens (1989), p. 20.

62. Riemens (1989), p. 72. Alongside conference co-organizers, Patrice Riemens and Rop Gonggrijp, Caroline Nevejan (2007) registered that the gathering had a double agenda, so to speak. It was not meant to be only a party, but to be tasked with drafting a declaration, which they did, representing an early hacker engagement with public policy that would reappear in the mid-1990s and then again with full force in the 2000s with the rise of hacker activism. For more details about this type of political engagement on an international scale, see Maxigas (2012); Coleman (2016); and Menn (2019).

63. Goerzen and Coleman (2022).

64. See Coleman (2015); Follis and Fish (2020); and Zetter (2014) for the description of these opposed trajectories.

65. Dardot and Laval (2014), p. 163.

66. Simondon (2014), p. 402.

Chapter 1

1. Cohn (2014).

2. See Turner (2008) and Lallement (2015) for a historical study of this legacy in the San Francisco Bay Area.

3. Walker (2018).

4. The Arduino project was founded in 2005 and became one of the most popular platforms for electronic prototyping in community spaces and educational projects worldwide. As far as its origin story goes, the founders of the project were working for the Interaction Design Institute Ivrea (IDII) in Ivrea, Italy, when the school lost its source of funding and was forced to shut down. To prevent the Arduino project from being inventoried as school property, the project members decided to license it as Open Hardware. At the time, one of their design students, Hernando Barragán, contributed an important addition to the project, Wiring, a programming environment for the Arduino project based on Processing, another popular programming environment for artists and designers. The basic impulse for creating the Arduino project came out of the need to have practical tools to teach students to create interactive design projects with basic electronics. The Arduino founders named the project after a pub in Ivrea, where they gathered regularly.

5. Foucault's definition of *heterotopic space* is particularly important for our discussion of community spaces. According to him, heterotopia consists of a "set of relations which delineates sites that are irreducible to one another," pointing to the specificity of socioeconomic and political forces intersecting local, regional, and transnational dimensions in emergent localities. Hackerspaces can be understood as heterotopic places since they stand as "singular spaces to be found in some given social spaces whose functions are different or the opposite of [institutionalized] others" (Foucault 1984, p. 252).

6. See Ingold (2000); Steil and Carvalho (2014); Steil, Carvalho, and Gomes

(2015); and Lave (2011) for the anthropological treatment of situated learning.

7. On deutero-learning, see Bateson (2000) and discussion in the Introduction, note 57.

8. Lave (2019).

9. This phenomenon has been conceptualized in terms of "collective effervescence" by Émile Durkheim ([1912] 1990). In the context of social movements, an updated version of the concept can be found in Victor Turner's (1969) description of *communitas*. For a treatment of the experience of *communitas* in hacker conferences, see Coleman (2012).

10. LadyAda and Altman (2013), https://perma.cc/F3VR-43UY.

11. Mauss (1950).

12. Lallement (2015).

13. Levy (2010).

14. In system administration, "sudo" is a command on GNU/Linux and BSD systems for the execution of administrator ("su," super user) commands under a regular user account. It is employed colloquially as a metaphor for the practice of self-entitlement meant to bypass the need for special roles and privileges.

15. Noisebridge mailing list (2013), https://perma.cc/FGU4-3APT.

16. Lizzard (2012).

17. Noisebridge wiki (2010-11-09), https://perma.cc/RF86-R4XZ.

18. Gennep (1960); Turner (1969).

19. Graeber (2009); Maeckelbergh (2009, 2011); Appel (2014); and Juris (2012).

20. Maxigas (2012).

21. Freeman (1972).

22. O'Brien (2012).

Chapter 2

1. Idyll Dandy Art website (2012), https://perma.cc/E8FT-MHNG.

2. The first attempt to bring collective attention to the need for common specifications for hardware was proposed by Bruce Perens in 1995 under the title "Open Hardware Specification Definition." The project was not embraced by the community (or the industry) and was discontinued. The proposal for Open Design and Open Source Hardware would be organized later by engineers and artists in the context of the nascent hackerspace movement in Europe and the US. See Powell (2012) and Gibb (2015) for a historical reconstruction of this period.

3. For the historical and ethnographic study of these collectivist experiences in the context of the alterglobalization movement and how they intersected with hacker collectives, see Juris (2008) and Maxigas (2012).

4. Altman (2011).

5. Altman (2012), https://perma.cc/RN9X-Y5N5.

Chapter 3

1. In his study of economic anthropology, Yunxiang Yan (1996) argued for the importance of the emotional dimension of gift giving that has been conspicuously absent in the literature. His argument is that the study of the gift operates as an analytic window into the "core of Chinese culture," with a focus on cultural meanings and operative logics of gift giving. Among emic categories, *lǐwù* (gift) is considered a central one, the ritualized "thing," represented by two Chinese characters (礼物) to express the normative, cultural aspect of exchange practices. According to Yan, Chinese scholars have developed a framework to study gift exchange in China based on local vernaculars such as "*guanxi* (personal networks), *renqing* (moral norms and human feelings), *mianzi* (face), and *bao* (reciprocity)" (p. 15). In this literature, *guanxi* is a key notion that "involves not only instrumentality and rational calculation, but also sociability, morality, intentionality, and personal affection" (p. 88). It stands for a "power game" as well as a lifestyle. It has an economic function as well as a social support function. *Renqing* operates as an ethical system, where reciprocity (*bao*) assumes different cultural forms. *Renqing* is the force behind the gift, the social force for solidarity, sharing, and cultivation of personal connections. This notion points to the moral obligation that is part of everyday gift giving. In Yan's words, "Gift giving is, perhaps, the most common channel for expressing one's emotional responses" (p. 141).

2. Gorz was highly influenced by his encounters with Ivan Illich in the 1970s, which, in part, explains his interest in "open technologies" as promising socioeconomic alternatives. He reminds us that the first title of Illich's *Tools for Conviviality* was *Retooling Society*, and it circulated as a publication of the Centro Intercultural de Documentación in Cuernavaca, Mexico. An inspirational and organic figure of ecological Marxism, Gorz was very hopeful about the transformative potential of collaborative digital collectives, but he did not live, due to tragic circumstances, to witness what the "knowledge economy" would become with the rise of surveillance platform capitalism (Srnicek 2016; Zuboff 2019; Terranova 2022). "Knowledge economy" meant something quite different for Gorz at the turn of the century: "A true knowledge economy," he asserts, "would be an economy based on zero-cost exchange and pooled resources, in which knowledge would be treated as humanity's common property. To exploit knowledge and turn it into capital, the capitalist enterprise has to privatize it and render scarce, by private appropriation and copyrighting, what is potentially abundant and free." (Gorz 2010, p. 179).

3. See Serrano (2017) for the analysis of the Open Hardware initiative at CERN.

4. For the analysis of this project and its connection with a longer history of Numeric Control machinery, see Söderberg (2014). For the discussion of the political implications of RepRap in relation to the history of hacking, see Söderberg and Maxigas (2020).

5. Söderberg (2014).

6. For an in-depth study of individualization in China, see Yan (2009) and Hansen and Svarverud (2010).

7. In this critique of substantivist epistemologies, Simondon elaborated the concept of "associated milieu" to emphasize the primacy of relations, not individual things. The philosopher explains the concept in these terms: "It can be affirmed that the individualization of technical beings is the condition of technical progress. This individualization is possible due to the recurrence of causality in a milieu that the technical being creates and conditions around it, that, in turn, conditions the technical being. This milieu is, at the same time, technical and natural and can be named associated milieu" (Simondon 2012 [1958], p. 70; my translation). Simondon's motivation for creating the concept is best explained in the exegesis of Barthélémy (2011): "purported 'things,' easily considered by common sense as something that exists before entering a relationship, are not made but of relations. To understand it, it is sufficient to stop wanting to account for a single individual, and take into account equally its 'associated milieu'" (p. 49, my translation).

8. Lindtner (2020).

9. Hayden (2010); Ribeiro (2013).

10. For classical and contemporary references on the topic, see Fei (1992); Yan (1996); Ong and Nonini (1996). See also note 1 in this chapter.

11. Huang (2017); Lindtner and Li (2012).

12. Gregory (1982) is a key reference in this debate for having theorized the relationship between gifts and commodities with an important distinction: whereas the gift is a measure of the person, the commodity is assumed to be an objective relation of equivalence between things or a measure, so to speak, of things. This debate was recuperated more recently by Tsing (2013) for the study of global production and circulation of gifts/commodities with the capacity to shift registers as they enter intricate supply chain networks.

13. Tsing's (2009) concept of supply chain capitalism is useful to illuminate Open Hardware businesses because their practices involve aligning contractors, subcontractors, volunteers, developers, designers, and partnerships with non-profit enterprises. In contrast with corporate supply chain enterprises of transnational scope, the nature of the Open Hardware supply chain involves distributed and independent research and development, shared learning resources online and offline, and self-organized collaborative spaces for testing, documenting, and using or improving designs at hackerspaces. Tsing (2009) makes an important distinction when defining the concept: "What is new is the hype and sense of possibility that supply chains offer to the current generation of entrepreneurs. This excitement creates its own effects, including the cascading rush toward outsourcing that has characterized the last two decades. This rush toward outsourcing, in turn, relies on new technologies that make it simple to communicate at a distance and send commodities with reliable speed.

It depends on new financial arrangements that make it easy to move money around and on new regimes of property that guarantee global profits. It is pressured by stockholder expectations for short-term corporate returns. It uses the enhanced mobility of labor and the economic and political vulnerabilities created by recent forms of imperialism and histories of global war. It sponsors new forms of 'creative' accounting and the auditing of 'immaterial' value. I call the subcontracting possibilities that elicit this new phenomenon of excitement 'supply chain capitalism'" (p. 149).

14. Weiner (1992).

15. Huang (2009).

16. Alford (1995); Ocko (1996); Yu (2012).

17. Hayden (2010).

18. Chan (2013); Irani (2019); Beltrán (2023); McElroy (2024).

19. Anderson (2012).

20. I follow Lisa Rofel (2017) in her depiction of cosmopolitanism as a "site of production of knowledge about what it means to be human in this reconfigured [neoliberal] world, knowledge that is embraced, digested, reworked, contested, and resisted in China" (p. 112).

21. Derrida's interpretation of the myth can be found in several book passages and recorded interviews. For a synthetic treatment of the allegory of Echo, see DeArmitt's (2009) "Resonances of Echo: A Derridean Allegory." I adopt DeArmitt's interpretation of the inventive potential in Echo's repetition to reflect on the excess that is produced through "copying" in the context of open projects across hackerspaces in the Global North and South: "What Derrida hears in Echo's reply to Narcissus is by no means an empty reduplication or a hollow reverberation of the same, but a unique and inventive response" (p. 95).

Chapter 4

1. For an anthropological treatment of the crisis of the globalized formal educational system and its shortcomings when it comes to effective, meaningful, and fulfilling learning, see Blum (2016).

2. Abelson, Sussman, and Sussman (1985).

3. The suggestion that "hackers are born, not made" is a persistent cultural motif in computerdom. See Ensmenger (2010) for the historical analysis of this discourse.

4. Danfei and Nalaginrut (2014), https://perma.cc/B42E-C8TP.

5. Danfei and Nalaginrut (2014), https://perma.cc/B42E-C8TP.

6. See Godbout (2003) for the discussion of the phenomenon of gift giving among strangers.

7. Jonas (1984).

Chapter 5

1. National Diet of Japan (2012); Shineha and Tanaka (2014); Cleveland, Knowles, and Shineha (2021).

2. Tanaka (2012); Nogrady (2023).

3. See Fortun (2001) and Masco (2006) as well as Illich (1973) on the topic of "technocratic disasters."

4. The work of Erving Goffman (1959) is a classical reference in micro-sociology for offering a lasting model for the study of social dramas. According to Goffman, "social establishments" can be analyzed from at least five different angles, which are all combined in the subjective experience of a sociotechnical circuit: *technical*, organized around efficiency as a system of activity; *political*, studied in terms of "the actions each participant can demand from other partic-ipants"; *structural*, in terms of the "horizontal and vertical status divisions and kinds of social relations"; *cultural*, in terms of the moral values which influence activity; and, finally, *dramaturgical*, a more holistic concept that intersects with all the previous four frames (p. 240). The dramaturgical angle concerns us the most here for having to do with what Goffman calls "impression management."

5. According to Coleman (2016), technical humor is a key characteristic of hacker sociality: "Hackers inject humor into every social situation and artifact. There is a long tradition of inserting small snippets of wit into code and docu-mentation; and they even embed hidden puzzles (what they call Easter eggs) in code for the amusement of those scrutinizing their work." (p. S93). For an extensive treatment of this aspect of hackerdom, see Coleman (2012). See also Thomas (2002) on the topics of "boy culture" and pranksterism in the so-called hacker underground.

6. BarCamps are independently organized gatherings for the discussion of digital technology topics. The name originates from the hacker slang for variable names "foo" and "bar." Foo Camp is an invitation-only conference organized by the publishing company O'Reilly Media. It is informally known as the Friends of O'Reilly Camp in reference to Timothy O'Reilly, Open Source evangelist and founder of O'Reilly Media. BarCamps follow the "unconference" format—that is, all the discussion topics and all the presentations are decided upon during the first hour of the event. The format is informal and flexible, accommodating var-ious forms of interaction from demonstrations to parallel conversation to more organized, structured talks. Participants are encouraged to present on a certain topic or offer a workshop instead of being mere spectators. Since its inception, it has become a very popular outside the United States and has ignited a network of events around the world.

7. *Otaku* is the Japanese term for someone who is obsessed with hobbies, which may involve computers, manga, anime, and video games. *Geek* would be the approximate synonym in English. For anthropological studies of otaku and geek cultures, see Allison (2006) and Kelty (2005), respectively.

8. Premeaux and Evans (2012).

9. The list of probing and humorous questions is too extensive to reproduce here, but they have the same structure and spirit, involving inquiries about basic technical topics that are assumed to be common knowledge in the community. TLS stands for Transport Layer Security; it is a very important protocol that guarantees secure (encrypted) communications for transacting machines on the internet.

10. For discussions of feminist hackerspaces as technopolitical alternatives, see Henry (2014); Toupin (2014); Savic and Wuschitz (2018); Dunbar-Hester (2020); and Richterich (2022).

11. Davies (2017), p. 105.

12. Drawing from Susan J. Douglas's work on technical cultures of masculinity, Dunbar-Hester (2014) described how (geek) radio activists, for example, fought against gender divisions and hierarchies in their collectives. In the authors' evaluation: "The historical legacy of radio and electronics as a masculine pastime and the gendered differences in individuals' personal backgrounds with regard to technical skill contributed to a gender gap" (Dunbar-Hester 2014, p. 54). This discussion is also presented (and treated more extensively) in Dunbar-Hester's 2020 book dedicated to the question of diversity in hackerspaces and open technology projects.

13. MRE (2011). Tokyo Hacker Space website, https://web.archive.org/web/20130316091934/http://www.tokyohackerspace.org/en/japan-in-crisis.

14. See Kimura (2016) and Sternsdorff-Cisterna (2018) for ethnographic studies of community-led food testing and safety networks and the relations they reconfigured with scientific expertise and governmental organizations after the Fukushima Daiichi nuclear meltdown.

15. See Morimoto (2022) on the kinship and spiritual relationality disrupted by the Fukushima disaster and by post-disaster public policy.

16. Rutherford and Geiger (1908).

17. Onda et al. (2020).

18. Knowles (2021) and Suzuki (2021).

19. See Kingston (2012) and Kimura (2016) for a critical analysis of the "nuclear village."

20. I presented this ethnographic mapping technique involving institutions, people, and technical objects in Murillo (2016).

21. Negri and Hardt (2019), p. 97.

22. Tokyo Hacker Space mailing-list, May 8, 2011, https://perma.cc/6GEJ-Y6X7.

23. Traweek (1988).

24. In his review of the metaphor of the bricoleur, Christopher Johnson (2012) exposes the backdrop against which the (gendered) figure of the scientist was elaborated in *The Savage Mind*: the nuclear physicist of the postwar period and the moral critique of the ethical implications of his scientific pursuit. If the scientist is Promethean, the author suggests, seeking to solve mys-

teries and advance knowledge of nature, the bricoleur is rather Epimethean, working creatively around the limits of his familiar surroundings. The practical knowledge of the engineer is supposed to be open and "projective"—that is, "it is able to invent its future without reference to its instrumental past," whereas for the bricoleur tools and materials "are over-determined by history of their usage [and] under-determined as for their potential use" (Johnson 2012, p. 362). Through ethnographic engagement, we get to see how hackers incorporate both the matter-of-fact sensibility of bricoleurs and, to a certain extent, the spiritual discipline of engineers. It is this ambiguity that makes hacking a critical practice with the *potential* to bring about alternative technological futures.

25. Lévi-Strauss ([1962] 2021), p. 25.

26. Lévi-Strauss ([1962] 2021), p. 40.

27. The interdisciplinary field of science and technology studies (STS) has contributed to this debate by complicating the Lévi-Straussean binary between the bricoleur and the engineer. In his discussion of bricolage, Donald MacKenzie (2003) examines the performative dimension of economic theory and finance. Through an analysis of the appropriations, detours, and tinkering in the history of option pricing theory, the author demonstrates the ad hoc and unpremeditated course of action for actors in the financial sector who are involved in theoretical work. Similarly, under the rubric of "social bricolage," Mario Biagioli (1992) demonstrates how the success of emergent scientific disciplines in late seventeenth-century Italy depended on the patrons' legitimacy, the establishment of scientific institutions, and the prestige of their locations. In a general sense, the concept of social bricolage circumscribes and identifies this particular political, cultural, and economic formation in which new scientific knowledge practices were made possible. Both McKenzie and Biagioli offer a renewed orientation for the study of emergent technoscientific fields by historicizing bricoleur practices, concepts, and artifacts.

28. Turkle and Papert (1990), p. 153.

29. Turkle and Papert (1990), p. 141.

30. For the examination of this important distinction between "reciprocal" and "permissive" Free and Open Source licensing—that is, between (reciprocal) licenses that require the use of the compatible licensing terms for derivative versions versus the (permissive) ones that allow for derivative versions to adopt restrictive, closed-source licensing terms—see Kelty (2008); Xifaras (2012); Coleman (2012); Vieira and de Filippi (2014).

31. Tanigaki et al (2015).

32. Benkler (2013).

33. Haraway (2016), p. 20.

34. Downey and Dumit (1997).

35. Kimura (2016).

36. Zigon (2018).

37. This concept is particularly helpful for studying blind spots and irrespon-

sibilities in technoscientific projects. Donna Haraway (2016) proposed an approach to response-ability that also included more-than-humans in relational arrangements.

38. See Juraku (2013) for a critical scrutiny of this report as "obscurantist" and "self-orientalist."

39. Bonner (2011).

40. See Downey (1986); Ogawa (2013); and Novak (2017).

41. See Dreyfus and Assange (1997) for details about this historic case.

42. Safecast-Japan mailing list, May 28, 2011, https://perma.cc/VNP9-9UC3.

43. Bonner (2011).

Chapter 6

1. Suetomi et al. (2021).

2. See Murillo and Bonner (2021) for the discussion of the scale of radioactive exposure adopted by Safecast.

3. DoseNet, University of California, Berkeley Department of Nuclear Engineering, https://radwatch.berkeley.edu/dosenet.

4. Kondo (1990).

5. The historical connection between computing and power stems from the use of human, electromechanical, and digital computers in war efforts, as discussed by Edwards (1996). The implications of this history at a micro sociological level for the persistent gender divides in engineering have also been explored by Edwards: "The equation between engineering, hardness, and masculinity, though very real in its exclusionary effects, tends to conceal the fact that the hard/soft split reappears within engineering at every level. In computer science the line can be drawn between software engineers, who program according to a disciplined, structural developmental logic, and hackers, undisciplined, often fanatical computer wizards whose nonstandard methods emerge from profound personal involvements with the machine" (Edwards 1990, p. 110).

6. Pfaffenberger (1996).

7. See Kelty (2008) for a description of the internal debate of the Linux development community regarding the proprietary code versioning software Bitkeeper. For the history of version control software in Silicon Valley, see Alcaras (2020).

8. Strathern (1988).

9. See Goerzen and Coleman (2022) for details of this formative conflict in the history of hacking.

10. *Tivoization* refers to the controversial case of TiVo Inc., a company that relied on Linux to run its digital video recorder but implemented a hardware restriction to assure that only software code created and signed by the company could run on the device. To prevent this practice, a new version of the General Public License (v3) was drafted with an "anti-Tivoization" clause. Hacker-turned-lawyer Bradley Kuhn describes the issue in the following terms: "Two

schools of thought about the purpose of copyleft have been at odds for some time. Simply put, the question is: are copyleft licenses designed primarily to protect the rights of large companies that produce electronics and software products, *or* is copyleft designed primarily to protect individual users' rights to improve, modify, repair, and reinstall their software?" (Kuhn 2021, https://perma.cc/ZE8Z-SXLS.

11. From the original text in the manifesto: "I consider that the golden rule requires that if I like a program I must share it with other people who like it" (Stallman 1985).

Conclusion

1. Mauss (1950).

2. Star and Bowker (2002).

3. Schmidt and Rosenberg (2014).

4. In her analysis of neoliberal governance, Wendy Brown (2016) drew from Foucault and Marx to argue that "emerging sharing or open source economies may be at odds with neoliberal governance seeding competition among radically decentralized units, but this governance modality introduces a brutal survivalism (and depressed remunerative scale) into these economies" (pp. 17–18). The misnomer "sharing economy," however, is not to be conflated with the ethnographic cases I discussed here—that is, with the Open Hardware production intra and inter sites, for example, with concrete alternatives of community-building that are set against growing individualization and IP enclosures. Despite the existence of the linkage that some of its members will make with startup and entrepreneurial ventures, I hope to have demonstrated that the hackerspace network is not the space of what Brown (2015) calls "economization," in which neoliberalism has become an all-encompassing economic rationality that colonizes many, if not most, domains of everyday life.

5. Vega, Lins Ribeiro, and Matthews (2015).

6. Brown (2016), p. 20.

7. The World Social Forum constituted an internationalist heterotopia where "another world" was declared to be "possible," according to various perspectives and in various languages. I use the term here in reference to a more subterranean tradition of the New Left. It is in the context of the autonomist tradition post–May 1968 that one finds the experimental politics of the "impossible" translated into a problem of political anthropology by Clastres (1974) to rethink the nature of political power beyond Euro-American traditions in political philosophy. It is also in this particular sense that I speak of an "impossible object" of inquiry to approach the *technopolitical otherwise* of alterglobalized computing, signaling the possibility of this alternative in the present. As I have discussed elsewhere (Murillo 2020b), my argument is not that hackerspaces are part of the anti-corporate globalization movement but that historic experiences with

convivial, autonomist, self-organized technopolitics find common ground in the latter and in the former.

8. Berrebi-Hoffmann, Bureau, Lallement (2018).

9. See Dunbar-Hester (2014); Nafus (2012) and Reagle (2013).

10. In the context of free radio activism in the United States, Christina Dunbar-Hester (2014) found the same issue: "More than one person told me that although there weren't any in their local group, women with skills were known to be 'out there.' These women attained a sort of mythic status, described as exceptionally 'hardcore' or 'kickass.' It was curious that the activists felt compelled to mention these skilled women; it nearly seemed like an apology for the disparity in their own group" (p. 55).

Bibliography

Abelson, Harold, Gerald Jay Sussman, and Julie Sussman. 1985. *Structure and Interpretation of Computer Programs*. New York: McGraw-Hill.

Alcaras, Gabriel. 2020. "Des biens industriels publics: Genèse de l'insertion des logiciels libres dans la Silicon Valley." *Sociologie du Travail* 62 (3).

Alford, William. 1995. *To Steal a Book Is an Elegant Offense: Intellectual Property Law in Chinese Civilization*. Stanford: Stanford University Press.

Allison, Anne. 2006. *Millennial Monsters: Japanese Toys and the Global Imagination*. Berkeley: University of California Press.

Altman, Mitch. 2012. "Hacking at the Crossroad: US Military Funding of Hackerspaces." *Journal of Peer Production* 2: 1–6.

Amrute, Sareeta. 2016. *Encoding Race, Encoding Class: Indian IT Workers in Berlin*. Durham: Duke University Press.

Anderson, Chris. 2012. *Makers: The New Industrial Revolution*. New York: Crown Business.

Apgaua, Renata. 2004. "O Linux e a perspectiva da dádiva." *Horizontes Antropológicos* 10 (21): 221–40.

Appel, Hannah. 2014. "Occupy Wall Street and the Economic Imagination." *Cultural Anthropology* 29 (4): 602–25.

Appadurai, Arjun. 1996. *Modernity at Large: Cultural Dimensions of Globalization*. Minneapolis: University of Minnesota Press.

Aspe, Bernard. 2002. "Être singulier commun." In Pascal Chabot, ed., *Simondon*. Paris: Vrin.

Augé, Marc. 1995. *Non-places: Introduction to an Anthropology of Supermodernity.* London; New York: Verso.

Bach, Jonathan. 2010. "'They Come in Peasants and Leave Citizens: Urban Villages and the Making of Shenzhen, China." *Cultural Anthropology* 25 (3): 421–58.

Barbrook, Richard, and Andy Cameron. 1996. "The Californian Ideology." *Science as Culture* 6 (1): 44–72.

Barthélémy, Jean-Hugues. 2011. *Simondon.* Paris: Les Belles Lettres.

Bateson, Gregory. 2000. *Steps to an Ecology of Mind.* Chicago: University of Chicago Press.

Beltrán, Héctor. 2023. *Code Work: Hacking across the US/México Techno-Borderlands.* Princeton: Princeton University Press.

Benkler, Yochai. 2006. *The Wealth of Networks: How Social Production Transforms Markets and Freedom.* New Haven: Yale University Press.

———. 2013. "Practical Anarchism: Peer Mutualism, Market Power, and the Fallible State." *Politics & Society* 41: 213–51.

Benkler, Yochai, and Helen Nissenbaum. 2006. "Commons-Based Peer Production and Virtue." *Journal of Political Philosophy* 14 (4): 394–419.

Berrebi-Hoffmann, Isabelle, Marie-Christine Bureau, and Michel Lallement. 2018. *Makers: Enquête sur les laboratoires du changement social.* Paris: Editions du Seuil.

Bessen, James, and Michael James Meurer. 2008. *Patent Failure: How Judges, Bureaucrats, and Lawyers Put Innovators at Risk.* Princeton: Princeton University Press.

Bessen, James E., Michael J. Meurer, and Jennifer Laurissa Ford. 2011. "The Private and Social Costs of Patent Trolls." Boston University School of Law, Law and Economics Research Paper No. 11-45. SSRN: https://dx.doi.org/10.2139/ssrn.1930272.

Beyer, Kurt. 2009. *Grace Hopper and the Invention of the Information Age.* Cambridge, MA: MIT Press.

Biagioli, Mario. 1992. "Scientific Revolution, Social Bricolage, and Etiquette." In Roy Porter and Mikuláš Teich, eds., *The Scientific Revolution in National Context*, 11–54. Cambridge: Cambridge University Press.

Blum, Susan D. 2016. *I Love Learning, I Hate School: An Anthropology of College.* Ithaca: Cornell University Press.

Boyer, Adrian. 2004. "Wealth Without Money." RepRap wiki, https://perma.cc/AW5F-SYDV.

Bonner, Sean. 2011. "Safecasting with Keio University." Safecast website, https://perma.cc/P269-J5YQ.

Brown, Wendy. 2015. *Undoing the Demos: Neoliberalism's Stealth Revolution.* Cambridge, MA: Zone Books.

———. 2016. "Power Visible and Invisible: Neoliberalism in Marxist and Foucauldian Frames." Presentation at the European Graduate School, Division of Philosophy, Art and Critical Thought, August 13.

Caillé, Alain. 2007. *Anthropologie du don: Le tiers paradigme*. Paris: La Découverte.

Campbell-Kelly, Martin, William Aspray, Nathan Ensmenger, and Jeffrey R. Yost. 2019. *Computer: A History of the Information Machine*. 3rd ed. New York: Routledge.

Campos, Haroldo de. 1992. *Metalinguagem e outras metas: Ensaios de teoria e crítica literária*. São Paulo: Editora Perspectiva.

Cardoso, Márcia de Oliveira. 2003. *O Patinho Feio como construção sociotécnica*. PhD diss., UFRJ/NCE/IM, Rio de Janeiro.

Chan, Anita. 2013. *Networking Peripheries: Technological Futures and the Myth of Digital Universalism*. Cambridge, MA: MIT Press.

Clastres, Pierre. 1974. *La société contre l'État: Recherches d'anthropologie politique*. Paris: Éditions de minuit.

Cleveland, Kyle, Scott Gabriel Knowles, and Ryuma Shineha. 2021. *Legacies of Fukushima: 3-11 in Context*. Philadelphia: University of Pennsylvania Press.

Cohn, Jesse. 2014. *Underground Passages: Anarchist Resistance Culture*. Oakland: AK Press.

Coleman, Gabriella. 2012. *Coding Freedom: The Ethics and Aesthetics of Hacking*. Princeton: Princeton University Press.

——. 2015. *Hacker, Hoaxer, Whistleblower, Spy: The Many Faces of Anonymous*. London: Verso.

——. 2016. "From Internet Farming to Weapons of the Geek." *Current Anthropology* 58 (S15): S91–102.

Coleman, E. Gabriella, and Alex Golub. 2008. "Hacker Practice." *Anthropological Theory* 8 (3): 255–77.

Dardot, Pierre, and Christian Laval. 2014. *Commun: Essai sur la Révolution au XXIe Siècle*. Paris: La Découverte.

Davies, Sarah. 2017. *Hackerspaces: Making the Maker Movement*. London: Wiley.

DeArmitt, Pleshette. 2009. "Resonances of Echo: A Derridean Allegory." *Mosaic: An Interdisciplinary Critical Journal* 42 (2): 89–100.

Derrida, Jacques. 1978. *Writing and Difference*. Chicago: University of Chicago Press.

Downey, Gary Lee. 1986. "Risk in Culture: The American Conflict over Nuclear Power." *Cultural Anthropology* 1 (4): 388–412.

Downey, Gary Lee, and Joseph Dumit. 1997. *Cyborgs and Citadels: Anthropological Interventions in Emerging Sciences and Technologies*. School of American Research Advanced Seminar Series. Santa Fe, NM: School of American Research Press.

Downey, Gary Lee, Joseph Dumit, and Sarah Williams. 1995. "Cyborg Anthropology." *Cultural Anthropology* 10 (2): 264–69.

Drahos, Peter, and Ruth Mayne. 2002. *Global Intellectual Property Rights: Knowledge, Access, and Development*. Oxford: Palgrave Macmillan.

Dreyfus, Suelette. 1997. *Underground: Tales of Hacking, Madness, and Obsession on the Electronic Frontier*. London: Mandarin.

Driscoll, Kevin. 2022. *The Modem World: A Prehistory of Social Media*. New Haven: Yale University Press.

Dunbar-Hester, Christina. 2014. *Low Power to the People: Pirates, Protest, and Politics in FM Radio Activism*. Cambridge, MA: MIT Press.

———. 2020. *Hacking Diversity: The Politics of Inclusion in Open Technology Cultures*. Princeton: Princeton University Press.

Durkheim, Émile. (1912) 1990. *Les formes élémentaires de la vie religieuse*. Paris: Presses Universitaires de France.

Edwards, Paul. 1990. "The Army and the Microworld: Computers and the Politics of Gender Identity." *Signs* 16 (1): 102–27.

———. 1996. *The Closed World: Computers and the Politics of Discourse in Cold War America*. Cambridge, MA: MIT Press.

Ensmenger, Nathan. 2010. *The Computer Boys Take Over: Computers, Programmers, and the Politics of Technical Expertise*. Cambridge, MA: MIT Press.

———. 2015. "'Beards, Sandals, and Other Signs of Rugged Individualism': Masculine Culture within the Computing Professions." *Osiris* 30 (1): 38–65.

Escobar, Arturo. 1994. "Welcome to Cyberia: Notes on the Anthropology of Cyberculture." *Current Anthropology* 35 (3): 211–31.

———. 2018. *Designs for the Pluriverse: Radical Interdependence, Autonomy, and the Making of Worlds*. Durham: Duke University Press.

Eve, Martin Paul. 2021. *Warez: The Infrastructure and Aesthetics of Piracy*. London: Punctum Books.

Farmer, Dan, and Wietse Venema. 1993. "Improving the Security of Your Site by Breaking Into It." Unpublished technical report. https://perma.cc/XT9Y -GGS6.

Federici, Silvia. 2018. *Re-enchanting the World: Feminism and the Politics of the Commons*. Toronto: PM Press.

Fei, Xiaotong. 1992. *From the Soil, the Foundations of Chinese Society: A Translation of Fei Xiaotong's Xiangtu Zhongguo*. Berkeley: University of California Press.

Fischer, Michael M. J. 2003. *Emergent Forms of Life and the Anthropological Voice*. Durham: Duke University Press.

———. 2009. *Anthropological Futures*. Durham: Duke University Press.

Follis, Luca and Adam Fish. 2020. *Hacker States*. Cambridge, MA: MIT Press.

Fortun, Kim. 2001. *Advocacy after Bhopal: Environmentalism, Disaster, New Global Orders*. Chicago: University of Chicago Press.

Foucault, Michel. 1984. *The Foucault Reader*. Edited by Paul Rabinow. New York: Pantheon Books.

———. 1988. *Technologies of the Self: A Seminar with Michel Foucault*. Amherst: University of Massachusetts Press.

Freeman, Jo. 1972. "The Tyranny of Structurelessness." *Berkeley Journal of Sociology* 17: 151–64.

Fritz, W. Barkley. 1996. "The Women of ENIAC." *IEEE Annals of the History of Computing* 18 (3): 13–28.

Gabrys, Jennifer. 2013. *Digital Rubbish: A Natural History of Electronics*. Ann Arbor: University of Michigan Press.

Gell, Alfred. 1992. "The Technology of Enchantment and the Enchantment of Technology." In Jeremy Coote and Anthony Shelton, eds., *Anthropology, Art and Aesthetics*, 40–63. Oxford: Clarendon Press.

Gennep, Arnold van. 1960. *The Rites of Passage*. Chicago: University of Chicago Press.

Gibb, Alicia. 2015. *Building Open Source Hardware: DIY Manufacturing for Hackers and Makers*. Upper Saddle River, NJ: Addison-Wesley.

Godbout, Jacques. 2003. *The World of the Gift*. Montreal: McGill-Queen's University Press.

Goerzen, Matt, and Gabriella Coleman. 2022. *Wearing Many Hats: The Rise of the Professional Security Hacker*. New York: Data & Society.

Goffman, Erving. 1959. *The Presentation of Self in Everyday Life*. New York: Doubleday.

Goldstein, Emmanuel. 2009. *Best of 2600: A Hacker Odyssey*. Hoboken, NJ: Wiley.

Gorz, André. 2010. *Ecologica*. New York: Seagull Books.

Graeber, David. 2009. *Direct Action: An Ethnography*. Oakland: AK Press.

Gregory, Chris. 1982. *Gifts and Commodities*. London; New York: Academic Press.

Gupta, Akhil. 1997. *Culture, Power, Place: Explorations in Critical Anthropology*. Durham: Duke University Press.

Gupta, Akhil, and James Ferguson. 1997. *Anthropological Locations : Boundaries and Grounds of a Field Science*. Berkeley: University of California Press.

Gusterson, Hugh. 1996. *Nuclear Rites: A Weapons Laboratory at the End of the Cold War*. Berkeley: University of California Press.

Hansen, Mette Halskov, and Rune Svarverud. 2010. *IChina: The Rise of the Individual in Modern Chinese Society*. Copenhagen: NIAS.

Haraway, Donna. 1991. *Simians, Cyborgs, and Women: The Reinvention of Nature*. New York : Routledge.

——. 2016. *Staying with the Trouble: Making Kin in the Chthulucene*. Durham: Duke University Press.

Harney, Stefano, and Fred Moten. 2013. *The Undercommons: Fugitive Planning and Black Study*. New York: Minor Compositions.

Hayden, Cori. 2010. "The Proper Copy." *Journal of Cultural Economy* 3 (1): 85–102.

Harvey, David. 2011. "The Future of the Commons." *Radical History Review* 2011 (109): 101–7.

Henry, Liz. 2014. "The Rise of Feminist Hackerspaces and How to Make Your Own." *Model View Culture* 2. https://perma.cc/LC6S-JPDN.

Hess, Charlotte, and Elinor Ostrom. 2007. *Understanding Knowledge as a Commons: From Theory to Practice*. Cambridge, MA: MIT Press.

Hicks, Marie. 2018. *Programmed Inequality*. Cambridge, MA: MIT Press.

Himanen, Pekka. 2001. *The Hacker Ethic and the Spirit of the Information Age*. New York: Random House.

Hoffman, Abbie. 1995. *Steal This Book*. New York: Four Walls Eight Windows.

Hopper, Grace. 1986. "Only the Limits of Our Imagination." *Chips Ahoy.* https://perma.cc/2GDX-X8UG.

Huang, Andrew. 2009. "Tech Trend: Shanzhai." *Bunnie's Blog.* https://perma.cc/XK69-7WCF.

———. 2017. *The Hardware Hacker: Adventures in Making and Breaking Hardware*. San Francisco: No Starch Press.

Illich, Ivan. 1971. *Deschooling Society*. New York: Harper & Row.

———. 1973. *Tools for Conviviality*. London: Boyars.

Ingold, Tim. 2000. *The Perception of the Environment: Essays on Livelihood, Dwelling and Skill*. London: Routledge.

Irani, Lily. 2019. *Chasing Innovation: Making Entrepreneurial Citizens in Modern India*. Princeton: Princeton University Press.

Johns, Adrian. 2009. *Piracy: The Intellectual Property Wars from Gutenberg to Gates*. Chicago: University of Chicago Press.

Johnson, Christopher. 2012. "Bricoleur and Bricolage: From Metaphor to Universal Concept." *Paragraph* 35 (3): 355–72.

Jonas, Hans. 1984. *The Imperative of Responsibility: In Search of an Ethics for the Technological Age*. Chicago: University of Chicago Press.

Juraku, Kohta. 2013. "'Made in Japan: A Critical Review for Accident Investigation Activities in Japan." In *STS Forum on East Japan Disaster*. https://perma.cc/8K4K-6PHM.

Juris, Jeffrey. 2008. *Networking Futures : The Movements against Corporate Globalization*. Durham: Duke University Press.

———. 2012. "Reflections on #Occupy Everywhere: Social Media, Public Space, and Emerging Logics of Aggregation." *American Ethnologist* 39 (2): 259–79.

Karsenti, Bruno. 1997. *L'homme total: Sociologie, anthropologie et philosophie chez Marcel Mauss*. Paris: Presses Universitaires de France.

Kelty, Christopher. 2005. "Geeks, Social Imaginaries, and Recursive Publics." *Cultural Anthropology* 20 (2): 185–214.

———. 2008. *Two Bits : The Cultural Significance of Free Software*. Durham: Duke University Press.

———. 2010. "Inventing Copyleft." In Mario Biagioli, Peter Jaszi, and Martha Woodmansee, eds., *Contexts of Invention*. Chicago: University of Chicago.

Kimura, Aya Hirata. 2016. *Radiation Brain Moms and Citizen Scientists: The Gender Politics of Food Contamination after Fukushima*. Durham: Duke University Press.

Kingston, Jeff. 2012. *Natural Disaster and Nuclear Crisis in Japan: Response and Recovery after Japan's 3/11*. London: Routledge.

Knowles, Scott Gabriel. 2021. "What Was Learned from 3.11?" In Scott Gabriel Knowles, Kyle Cleveland, and Ryuma Shineha, eds., *Legacies of Fukushima: 3.11 in Context*, 17–34. Philadelphia: University of Pennsylvania Press.

Kondo, Dorinne. 1990. *Crafting Selves: Power, Gender, and Discourses of Identity in a Japanese Workplace*. Chicago: University of Chicago Press.

Kuhn, Bradley. 2021. "'Tivoization' & Your Right to Install Under Copyleft & GPL." Conservancy Blog, Software Freedom Conservancy. https://perma.cc/ZE8Z-SXLS.

Lahire, Bernard. 2002. *Portraits sociologiques: Dispositions et variations individuelles*. Paris: Nathan.

Lallement, Michel. 2015. *L'âge du faire: Hacking, travail, anarchie*. Paris: Seuil.

Landreth, Bill. 1985. *Out of the Inner Circle: A Hacker's Guide to Computer Security*. Bellevue, WA: Microsoft Press.

Lapsley, Phil. 2013. *Exploding the Phone: The Untold Story of the Teenagers and Outlaws Who Hacked Ma Bell*. New York: Grove Press.

Lave, Jean. 2011. *Apprenticeship in Critical Ethnographic Practice*. Chicago: University of Chicago Press.

———. 2019. *Learning and Everyday Life: Access, Participation, and Changing Practice*. Cambridge: Cambridge University Press.

Leach, James. 2009. "Freedom Imagined: Morality and Aesthetics in Open Source Software Design." *Ethnos* 74 (1): 51–71.

Lécuyer, Christophe. 2006. *Making Silicon Valley: Innovation and the Growth of High Tech, 1930-1970*. Cambridge: MIT Press.

Lévi-Strauss, Claude. 1955. *Tristes tropiques*. Paris: Plon.

———. (1962) 2021. *The Savage Mind*. Chicago: University of Chicago Press.

Levy, Steven. 2010. *Hackers: Heroes of the Computer Revolution*. Sebastopol, CA: O'Reilly.

Light, Jennifer. 1999. "When Computers Were Women." *Technology and Culture* 40 (3): 455–83.

Lindtner, Silvia. 2020. *Prototype Nation: China and the Contested Promise of Innovation*. Princeton: Princeton University Press.

Lindtner, Silvia, and David Li. 2012. "Created in China: The Makings of China's Hackerspace Community." *Interactions* 19 (6): 18–22.

Lizzard. (2012). "Love @ First Hack" *Zine in Progress* (1): 3.

Lukes, Steven. 1990. *Individualism: Key Concepts in the Social Sciences*. Oxford: Blackwell.

King, Martin Luther, Jr. 1968. "Beyond Vietnam." Speech delivered at Riverside Church, New York, April 4, 1968. https://perma.cc/3PKJ-TEMS.

Maeckelbergh, M. E. 2009. *The Will of the Many: How the Alterglobalization Movement Is Changing the Face of Democracy*. London: Pluto Press.

———. 2011. "Doing Is Believing: Prefiguration as Strategic Practice in the Alterglobalization Movement." *Social Movement Studies* 10 (1): 1–20.

MacIntyre, Alasdair. 1984. *After Virtue: A Study in Moral Theory*. Notre Dame, IN: University of Notre Dame Press.

MacKenzie, Donald. 2003. "An Equation and Its Worlds: Bricolage, Exemplars,

Disunity and Performativity in Financial Economics." *Social Studies of Science* 33 (6): 831–68.

Marcus, George. 1995. "Ethnography in/of the World System: The Emergence of Multi-Sited Ethnography." *Annual Review of Anthropology* 24: 95–117.

Marcus, George, and Michael Fischer. 1986. *Anthropology as Cultural Critique : An Experimental Moment in the Human Sciences.* Chicago: University of Chicago Press.

Marques, Ivan da Costa. 2005. "Cloning Computers: From Rights of Possession to Rights of Creation." *Science as Culture* 14 (2): 139–60.

Masco, Joseph. 2006. *The Nuclear Borderlands: The Manhattan Project in Post–Cold War New Mexico.* Princeton: Princeton University Press.

Mattingly, Cheryl, and Jason Throop. 2018. "The Anthropology of Ethics and Morality." *Annual Review of Anthropology* 47 (1): 475–92.

Mauss, Marcel. (1923) 1950. *Sociologie et anthropologie.* Paris: Presses Universitaires de France.

———. (1938) 2012. *Techniques, technologie et civilisation.* Edited by Florence Weber and Nathan Schlanger. Paris: Presses Universitaires de France.

Maxigas. 2012. "Hacklabs and Hackerspaces: Tracing Two Genealogies." *Journal of Peer Production* 2: 1–10.

McElroy, Erin. 2024. *Silicon Valley Imperialism: Techno Fantasies and Frictions in Postsocialist Times.* Durham: Duke University Press.

McIlwain, Charlton D. 2020. *Black Software: The Internet and Racial Justice, from the AfroNet to Black Lives Matter.* Oxford: Oxford University Press.

Menn, Joseph. 2019. *Cult of the Dead Cow: How the Original Hacking Supergroup Might Just Save the World.* New York: Public Affairs.

Mitnick, Kevin. 2011. *Ghost in the Wires: My Adventures as the World's Most Wanted Hacker.* New York: Little, Brown.

Moody, Glyn. 2001. *The Rebel Code: The inside Story of Linus and the Open Source Revolution.* Cambridge, Mass.: Perseus Pub.

Morimoto, Ryo. 2022. "A Wild Boar Chase: Ecology of Harm and Half-Life Politics in Coastal Fukushima." *Cultural Anthropology* 37 (1): 69–98.

Murillo, Luis Felipe R. 2010. "Tecnologia, política e cultura na comunidade brasileira de software livre e de código aberto." In Ondina Fachel Leal and Rebeca Hennemann Vergara de Souza, eds., *Do regime de propriedade intelectual: estudos antropológicos,* 75–93. Porto Alegre, BR: Tomo Editorial.

———. 2016. "New Expert Eyes over Fukushima: Open Source Responses to the 3/11 Disaster in Japan." *Anthropological Quarterly* 89 (2): 399–429.

———. 2020a. "Magie et/comme hacking." *Revue du MAUSS permanente,* November 27, 2020. https://perma.cc/CTL3-T873.

———. 2020b. "Hackerspace Network: Prefiguring Technopolitical Futures?" *American Anthropologist* 122 (2): 207–21.

Murillo, Luis Felipe R., and Sean Bonner. 2021. "Building a Community-Based Platform for Radiation Monitoring after 3.11." In Kyle Cleveland, Scott Gabriel Knowles, and Ryuma Shineha, eds., *Legacies of Fukushima: 3–11 in Context*, 183–96. Philadelphia: University of Pennsylvania Press.

Murillo, Luis Felipe R., and Christopher Kelty. 2017. "Hackers and Hacking." In G. Koch, ed., *Digitisation: Theories and Concepts for Empirical Cultural Research*, 95–116. London: Routledge.

Nafus, Dawn. 2012. "'Patches Don't Have Gender': What Is Not Open in Open Source Software." *New Media & Society* 14: 669–83.

National Diet of Japan. 2012. *National Diet of Japan Fukushima Nuclear Accident Independent Investigation Commission Report*. Tokyo: National Diet of Japan Fukushima Nuclear Accident Independent Investigation Commission Report. https://perma.cc/A24V-FWKH.

Nevejan, Caroline. 2007. *Presence and the Design of Trust*. PhD diss., University of Amsterdam.

Nevejan, Caroline, and Alexander Badenoch. 2014. "How Amsterdam Invented the Internet: European Networks of Significance, 1980–1995." In Gerard Alberts and Ruth Oldenziel, eds., *Hacking Europe: From Computer Cultures to Demoscenes*, 189–217. London: Springer.

Negri, Antonio, and Michael Hardt. 2019. *Assembly*. Oxford: Oxford University Press.

Nissenbaum, Helen. 2004. "Hackers and the Contested Ontology of Cyberspace." *New Media & Society* 6 (2): 195–217.

Nogrady, Bianca. 2023. "Is Fukushima Wastewater Release Safe? What the Science Says." *Nature* 618 (7967): 894–95.

Novak, David. 2017. "Project Fukushima! Performativity and the Politics of Festival in Post-3/11 Japan." *Anthropological Quarterly* 90 (1): 225–53.

Obeng-Odoom, Franklin. 2021. *Commons in an Age of Uncertainty*. Toronto: Toronto University Press.

O'Brien, Danny. 2012. "Is Noisebridge an Anarchist Hackerspace?" *Zine in Progress* (3): x099.

Ocko, Jonathan. 1996. "Copying, Culture, and Control: Chinese Intellectual Property Law in Historical Context," *Yale Journal of Law & the Humanities* 8 (2), article 10.

Ogawa, Akihiro. 2013. "Young Precariat at the Forefront: Anti-Nuclear Rallies in Post-Fukushima Japan." *Inter-Asia Cultural Studies* 14 (2): 317–26.

Onda, Yuichi, Keisuke Taniguchi, Kazuya Yoshimura, Hiroaki Kato, Junko Takahashi, Yoshifumi Wakiyama, Frederic Coppin, and Hugh Smith. 2020. "Radionuclides from the Fukushima Daiichi Nuclear Power Plant in Terrestrial Systems." *Nature Reviews Earth & Environment* 1 (12): 644–60.

Ong, Aihwa, and Donald M. Nonini. 1996. *Ungrounded Empires: The Cultural Politics of Modern Chinese Transnationalism*. New York: Routledge.

Ostrom, Elinor. 1990. *Governing the Commons*. Cambridge: Cambridge University Press.

Ostrom, Vincent, and Elinor Ostrom. 1977. "Public Goods and Public Choices."

In Emanuel S. Savas, ed., *Alternatives for Delivering Public Services: Toward Improved Performance*, 7–49. Boulder, CO: Westview Press.

P.M. 1994. *Bolo'bolo*. New ed. New York: Semiotext(e).

Pellow, David, and Lisa Sun-Hee Park. 2002. The Silicon Valley of Dreams: Environmental Injustice, Immigrant Workers, and the High-Tech Global Economy. NYU Press.

Pfaffenberger, Bryan. 1988. "The Social Meaning of the Personal Computer: Or, Why the Personal Computer Revolution Was No Revolution." *Anthropological Quarterly* 61 (1): 39–47.

———. 1992. "Social Anthropology of Technology." *Annual Review of Anthropology* 21: 491–516.

———. 1996. "'If I Want It, It's OK': Usenet and the (Outer) Limits of Free Speech." *The Information Society* 12 (4): 365–86.

Pina-Cabral, João. 2024. "Person and Relation as Categories: Mauss' Legacy." *History and Anthropology* 35 (1): 170–88. https://doi.org/10.1080/02757206.2021.2019031.

Powell, Alison. 2012. "Democratizing Production through Open Source Knowledge: From Open Software to Open Hardware." *Media, Culture & Society* 34 (6): 691–708.

Premeaux, Emery, and Brian Evans. 2012. *Arduino Projects to Save the World*. New York: Apress.

Rankin, Joy Lisi. 2018. *A People's History of Computing in the United States*. Cambridge, MA: Harvard University Press.

Raymond, Eric. 1993. *The New Hacker's Dictionary*. Cambridge, MA: MIT Press.

———. 1999. *The Cathedral and the Bazaar: Musings on Linux and Open Source by an Accidental Revolutionary*. Sebastopol, CA: O'Reilly Press.

Reagle, Joseph. 2013. "'Free as in Sexist?' Free Culture and the Gender Gap." *First Monday* 18 (1). https://doi.org/10.5210/fm.v18i1.4291.

Ribeiro, Gustavo Lins. 1994. *The Condition of Transnationality*. Brasília: Departamento de Antropologia, Universidade de Brasília.

———. 2013. "What's in a Copy?" *Virtual Brazilian Anthropology* 10 (1) (October): 20–39.

Richterich, Annika. 2022. "Hackerspaces as Technofeminist Sites for Experiential Learning." *Learning, Media and Technology* 47 (1): 11–25.

Riemens, Patrice. 1989. *Galactic Hacker Party (ICATA '89)*. Conference report. https://perma.cc/LN3C-8FDQ.

Rofel, Lisa. 2017. *Desiring China: Experiments in Neoliberalism, Sexuality, and Public Culture*. Durham: Duke University Press.

Rutherford, Ernest, and Hans Geiger. 1908. "An Electrical Method of Counting the Number of α Particles from Radioactive Substances." *Proceedings of the Royal Society* 81 (546): 141–61.

Salus, Peter H. 1994. *A Quarter Century of UNIX*. Reading, MA: Addison-Wesley.

Savic, Selena, and Stefanie Wuschitz. 2018. "Feminist Hackerspace as a Place of Infrastructure Production." *Ada: A Journal of Gender, New Media, and Technology* (13) (January). https://doi.org/10.5399/uo/ada.2018.13.10.

Serrano, Javier. 2017. *Open Hardware and Collaboration.* Keynote speech, 11th International Workshop on Personal Computers and Particle Accelerator Controls, Campinas, Brazil, October 27.

Scheirer, Walter. 2024. *A History of Fake Things on the Internet.* Stanford: Stanford University Press.

Schmidt, Eric, and Jonathan Rosenberg. 2014. *How Google Works.* London: John Murray Press.

Scott, Jason. 2008. *One Last Time: Hacking/Phreaking History Primer.* Conference presentation, Hackers on Planet Earth, New York City, July 23.

Shineha, Ryuma, and Mikihito Tanaka. 2014. "Mind the Gap: 3.11 and the Information Vulnerable." *The Asia-Pacific Journal: Japan Focus* 12 (7), no. 4.

Simondon, Gilbert. (1958) 2012. *Du mode d'existence des objets techniques.* Paris: Aubier.

——. 2014. *Sur la technique.* Paris: Presses Universitaires de France.

——. 2017. *L'individuation à la lumière des notions de forme et d'information.* Paris: Editions Jérôme Millon.

Schlanger, Nathan, ed. 2012. *Techniques, technologie et civilisation.* By Marcel Mauss. Paris: PUF.

Sigaut, François. 2011. "La Formule de Mauss." Techniques & Culture n 54-55 (1): 357–67.

Söderberg, Johan. 2007. *Hacking Capitalism: The Free and Open Source Software Movement.* London: Routledge.

——. 2014. "Reproducing Wealth without Money, One 3D Printer at a Time: The Cunning of Instrumental Reason." *Journal of Peer Production* 1 (4) (Winter).

Söderberg, Johan, and Maxigas. 2022. *Resistance to the Current: The Dialectics of Hacking.* Cambridge, MA: MIT Press.

Srnicek, Nick. 2016. *Platform Capitalism.* London: Polity.

Stallman, Richard. 1985. "The GNU Manifesto." *Dr. Dobb's Journal of Software Tools* 10 (3): 30–32.

Stallman, Richard, and Larry Wall. 2002. "Sharing the Source." In Adam Brate, ed., *Technomanifestos: Visions from the Information Revolutionaries,* 240–68. New York: Texere.

Star, Susan Leigh, and Geoffrey Bowker. 2002. "How to Infrastructure." In L. A. Lievrouw and S. Livingstone, eds., *Handbook of New Media. Social Shaping and Consequences of ICTs,* 151–62. London: Sage.

Steil, Carlos Alberto, and Isabel Cristina de Moura Carvalho. 2014. "Epistemologias ecológicas: Delimitando um conceito." *Mana* 20 (1): 163–83.

Steil, Carlos Alberto, Isabel Cristina de Moura Carvalho, and Ana Maria R. Gomes. 2015. "Apresentação." *Horizontes Antropológicos* 21 (December): 9–17.

Sternsdorff-Cisterna, Nicolas. 2018. *Food Safety after Fukushima: Scientific Citizenship and the Politics of Risk*. Honolulu: University of Hawaii Press.

Sterling, Bruce. 1992. *The Hacker Crackdown: Law and Disorder on the Electronic Frontier*. New York: Bantam Books.

Strathern, Marilyn. 1988. *The Gender of the Gift: Problems with Women and Problems with Society in Melanesia*. Berkeley: University of California Press.

Stoll, Clifford. 1989. *The Cuckoo's Egg: Tracking a Spy through the Maze of Computer Espionage*. New York: Doubleday.

Suetomi, Eiichi, Yuko Hatano, Masakiyo Fujita, Yukiko Okada, Kyuma Suzuki, and Shun Watanabe. 2021. "Long-Term Prediction of 137 Cs in Lake Onuma on Mt. Akagi after the Fukushima Accident Using Fractional Diffusion Model." *Scientific Reports* 11 (1): 20336.

Suzuki, Tatsujiro. 2021. "Has Japan Learned a Lesson from the Fukushima Nuclear Accident?" In Kyle Cleveland, Scott Gabriel Knowles, and Ryuma Shineha, eds., *Legacies of Fukushima: 3.11 in Context*, 63–76. Philadelphia: University of Pennsylvania Press.

Tanaka, Shun-ichi. 2012. "Accident at the Fukushima Dai-Ichi Nuclear Power Stations of TEPCO—Outline and Lessons Learned." *Proceedings of the Japan Academy* 88 (9): 471–84.

Tanigaki, M., R. Okumura, K. Takamiya, N. Sato, H. Yoshino, H. Yoshinaga, Y. Kobayashi, A. Uehara, and H. Yamana. 2015. "Development of KURAMA-II and Its Operation in Fukushima." *Nuclear Instruments and Methods in Physics Research Section A: Accelerators, Spectrometers, Detectors and Associated Equipment* 781 (May): 57–64.

Taylor, Paul A. 1999. *Hackers: Crime and the Digital Sublime*. London: Taylor and Francis.

Tarot, Camille. 2003. *Sociologie et anthropologie de Marcel Mauss*. Paris: La Découverte.

Terranova, Tiziana. 2022. *After the Internet: Digital Networks between Capital and the Common*. Cambridge, MA: MIT Press.

Thomas, Douglas. 2002. *Hacker Culture*. Minneapolis: University of Minnesota Press.

Thompson, Ken. 1984. "Reflections on Trusting Trust." *Communications of the ACM* 27 (8): 761–63.

Torvalds, Linus. 2001. *Just for Fun: The Story of an Accidental Revolutionary*. New York: HarperBusiness.

Toupin, Sophie. 2014. "Feminist Hackerspaces: The Synthesis of Feminist and Hacker Cultures." *Journal of Peer Production* 5 (2014): 1–11.

Traweek, Sharon. 1988. *Beamtimes and Lifetimes: The World of High Energy Physicists*. Cambridge, MA: Harvard University Press.

Tsing, Anna. 2005. *Friction: An Ethnography of Global Connection*. Princeton: Princeton University Press.

———. 2009. "Supply Chains and the Human Condition." *Rethinking Marxism* 21 (2): 148–76.

———. 2013. "Sorting out Commodities: How Capitalist Value Is Made through Gifts." *HAU: Journal of Ethnographic Theory* 3 (1): 21–43.

Turkle, Sherry. 1984. *The Second Self: Computers and the Human Spirit*. New York: Simon and Schuster.

Turkle, Sherry, and Seymour Papert. 1990. "Epistemological Pluralism: Styles and Voices within the Computer Culture." *Signs: Journal of Women in Culture and Society* 16 (1): 128–57.

Turner, Fred. 2008. *From Counterculture to Cyberculture: Stewart Brand, the Whole Earth Network, and the Rise of Digital Utopianism*. Chicago: University of Chicago Press.

Turner, Victor. 1969. *The Ritual Process: Structure and Anti-Structure*. Chicago: University of Chicago Press.

Vega, Carlos Alba, Gustavo Lins Ribeiro, and Gordon Mathews. 2015. *La globalización desde abajo: La otra economía mundial*. Mexico DF: Fondo de Cultura Económica.

Vieira, Miguel Said, and Primavera de Filippi. 2014. "Between Copyleft and Copyfarleft: Advanced Reciprocity for the Commons." *Journal of Peer Production* no. 4: *Value and Currency*.

Viveiros de Castro, Eduardo. 2015. *Metafísicas Canibais*. São Paulo: Cosac Naify.

Walker, Richard. 2018. *Pictures of a Gone City: Tech and the Dark Side of Prosperity in the San Francisco Bay Area*. Oakland: PM Press.

Weber, Steven. 2004. *The Success of Open Source*. Cambridge, MA: Harvard University Press.

Weiner, Annette. 1992. *Inalienable Possessions: The Paradox of Keeping-While-Giving*. Berkeley: University of California Press.

Williams, Sam, and Richard Stallman. 2002. *Free as in Freedom. Richard Stallman's Crusade for Free Software*. Sebastopol, CA: O'Reilly Press.

Xifaras, Mikhail. 2012. "Copyleft and the Theory of Property." *Multitudes* 2010/2 (41): 50–64.

Yan, Yunxiang. 1996. *The Flow of Gifts: Reciprocity and Social Networks in a Chinese Village*. Stanford: Stanford University Press.

———. 2009. "The Good Samaritan's New Trouble: A Study of the Changing Moral Landscape in Contemporary China." *Social Anthropology* 17 (1): 9–24.

Yu, Peter. 2012. "Intellectual Property and Asian Values." *Marquette Intellectual Property Law Review* 16 (2): 329.

Zeitlyn, David. 2003. "Gift Economies in the Development of Open Source Software: Anthropological Reflections." *Research Policy, Open Source Software Development*, 32 (7): 1287–91.

Zetter, Kim. 2014. *Countdown to Zero Day: Stuxnet and the Launch of the World's First Digital Weapon*. New York: Crown.

Zigon, Jarrett. 2008. *Morality: An Anthropological Perspective*. New York: Berg.

——. 2011. *HIV Is God's Blessing: Rehabilitating Morality in Neoliberal Russia*. Berkeley: University of California Press.

——. 2018. "Can Machines Be Ethical? On the Necessity of Relational Ethics and Empathic Attunement for Data-Centric Technologies." *Social Research: An International Quarterly* 86 (4): 1001–22.

Zigon, Jarrett, and C. Jason Throop. 2014. "Moral Experience: Introduction." *Ethos* 42 (1): 1–15.

Zuboff, Shoshana. 2019. *The Age of Surveillance Capitalism: The Fight for a Human Future at the New Frontier of Power*. New York: Public Affairs.

Index

Abelson, Hal, 105, 124
Ada, Lady, 36
Adafruit, 92
Agile Manufacturing Center, 91–92, 93
agile software development, 152
Akagisan, Japan, 137
Alibaba, 100
Alonzo Church (Noisebridge), 30
alterglobalization of computing expertise, 20, 24–25
Altman, Mitch: on anarchy, 71; awards of, 72; background of, 61–63; on Boy Scouting, 61–62; at Circuit Hacking Monday, 33, 34; on education, 63; education of, 63–64; on flight simulation program, 65; on hackerspaces, 69; influence of, 72, 156, 157; at International Free Software Forum (FISL), 60; inventions of, 68; on love, 157; on Open Source, 60–61, 70, 73; on political organization, 64; on Rainbow Gatherings, 64–65, 67; socialist experiences of, 83; socialization of, 63; soldering and, 35; as teaching assistant, 63–64; 3ware of, 67–68; at Tokyo Hacker Space (THS), 115; travels of, 64, 66; TV-B-Gone of, 68–70, 73–74, 92; work of, 65–67, 68; writings of, 72
American Anthropological Association, 167n44
Anderson, Chris, 99
anthropology of techniques, 163n6
Anti-966 license, 101
Arduino project, 96, 97–99, 171n4
ARPANET, 10
Ars Electronica, 124
Artificial Intelligence Lab (MIT), 13
Aspray, William, 164n13
Assange, Julian, 166n31
associated milieu, 84, 174n7
Association for Citizens and Scientists Concerned about Internal Radiation Exposures, 129

The authorized representative in the EU for product safety and compliance is:
Mare Nostrum Group B.V.
Mauritskade 21D
1091 GC Amsterdam
The Netherlands
Email address: gpsr@mare-nostrum.co.uk

KVK chamber of commerce number: 96249943

The authorized representative in the EU for product safety and compliance is:
Mare Nostrum Group
B.V Doelen 72
4831 GR Breda
The Netherlands

www.ingramcontent.com/pod-product-compliance
Lightning Source LLC
Chambersburg PA
CBHW030818270326
41928CB00007B/782

* 9 7 8 1 5 0 3 6 4 1 4 8 8 *